An Introduction

To

Engineering Design

ISBN: 978-1-887503- 01-3

SDC
PUBLICATIONS
Schroff Development Corporation

www.schroff.com
www.schroff-europe.com

Many thanks to the following people without whose help this book could not have been compiled.

Dr. Robert Arnzen

Dr. Andrew Dimarogonas

Mr. Andrew Lindberg

Mr. Charles Muench

Mr. James Watson

Special acknowledgment goes to Mr. Stephen Schroff, the Publisher

This book is dedicated to Mr. Christopher Kimmel who on a daily basis must overcome difficulties that would discourage the best of us. Who has accomplished more than most of us. Whose good humor, indomitable spirit and ornery nature has carried him through.

Table of Contents

Notes:

Introduction to Engineering Design

Engineering Education has recently placed more emphasis on design in the curriculum.

Why is design an important component of Engineering education? <u>The American Heritage Dictionary</u> provides perspective.

en·gi·neer·ing (µn”j…-nîr“1ng) n. Abbr. e., E., eng. 1.a. The application of scientific and mathematical principles to practical ends such as the design, manufacture, and operation of efficient and economical structures, machines, processes, and systems. b. The profession of or the work performed by an engineer.

—**de·sign** n. 1.a. A drawing or sketch. b. A graphic representation, especially a detailed plan for construction or manufacture. 2. The purposeful or inventive arrangement of parts or details.

Engineering design is the activity that skillfully assembles fragments from many technical studies into useful products or processes. Engineering education is structured around these statements. Many courses at first broadly involve basic science and math. These courses sharpen the ability to observe, reason, measure and analyze. Later courses focus on narrower fields of study to provide depth in a specific curriculum. Machine design, circuit design, structural design ...., etc. introduce accepted standards found in each field. Projects are used to show how to approach a design situation, research existing knowledge and apply standards.

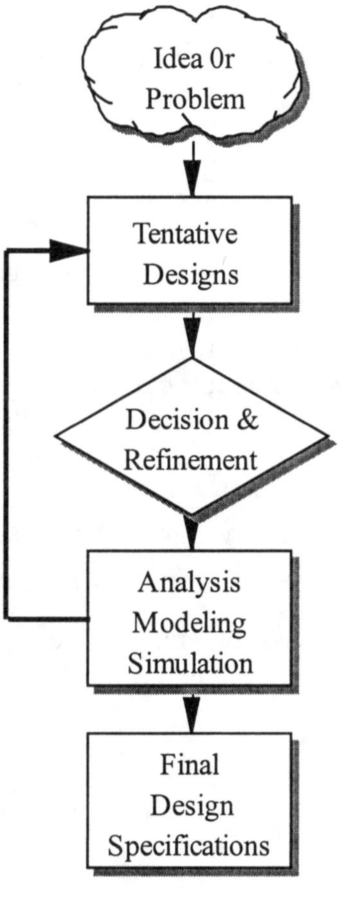

Classical Design Model

Many writers have researched the sequence of events depicting the design process. The block diagram to the left is a simplified representation. This model is changing. Concurrent Engineering, for example, transforms the linear design cycle into a more parallel design activity.

- A new idea or a problem with an existing product or process is the motivation to start the design process.
- Preliminary designs encourage innovation and creativity.
- Decisions must be made at every point in the design.
- Analysis and simulation verify the design.
- Final design conveys graphical and operational information. This information is used in the detailed design of the product.

"Design, which is now largely dedicated to superficial ends, is appropriate to our most significant human activities, and belongs to them." Ralph Kaplan in By Design.

How do new designs originate? Three general reasons — scientific discoveries, dis-satisfaction with a current product or a need for a new product. On rare occasions, like the discovery of the transistor or the laser or a synthetic fiber, new products evolve. Few people would go back to the pops, cracks and deterioration of music recorded on vinyl when compact disks provide consistent high quality. Product designs based on scientific discoveries are often very speculative and difficult to project to the marketplace.

Dissatisfaction with a current product may trigger a re-design. This dissatisfaction might be brought on by an outmoded appearance, economic competition, a new world market, ecological constraints, customer expectations, etc. Consumers have been conditioned to want constant "improvements" in automobiles, washing machines and computer software. The yearly upgrade is expected whether it is needed or not. Consumer driven product design may result in incorporation of trivial features. Word processors for computers have expanded into areas which few users need or ever use. The penalty is cost and resource overhead.

"Necessity equals invention." Charlie Muench, Chemical Engineer and Inventor.

Needs occur for products. Special items or systems are required for disabled persons. Machines, chemical processes or computer controls must be designed for the manufacturing industry. Once a product or system is defined, a whole design process kicks in to develop the production tools to manufacture components. Even a small piece may require major effort to produce due to geometry or mechanical requirements. The average consumer never sees or is ever aware of the work that goes on behind the fancy package. Boarding an airplane, the traveler is unaware of the incredibly detailed engineering required for each part to assure minimum weight and optimum strength.

Constraints.

Many outside forces must be reckoned. Sometimes these forces appear unreasonable or ridiculous. Especially those seemingly imposed by society or politics.

- Ecology.
- Safety.
- Energy Consumption.
- Maintainability.
- Zoning laws
- Political power
- "The boss said to do it this way"
- Cost, etc. ....

The Work of the Engineer

"We are delighted whenever someone makes what we wanted and never new we wanted." Ralph Kaplan in <u>By Design.</u>

An engineer designs something, then engineers it into reality. This work is driven by the needs of society. Engineers do not create need — they respond to it. Rarely does one person carry out all the jobs required. In actual practice teams of engineers work on projects. These people must understand a wide range of scientific, mathematical and humanistic phonomena.

Engineering students sometimes question the liberal education requirements. Why study literature when one wants to design computers? The answer is that most designs are for and about people. People must use and approve a product. Engineers must understand physical limitations, mental attitudes and predjuices. One vacuum cleaner manufacturer with a mediocre product line hired an outstanding woman engineer/designer to provide another perspective. Lower weight, less noise, better attachment design along with color changes resulted in increased sales and customer satisfaction.

Production /Consumption cycle. ref Morris Asimow <u>Introduction to Design.</u>

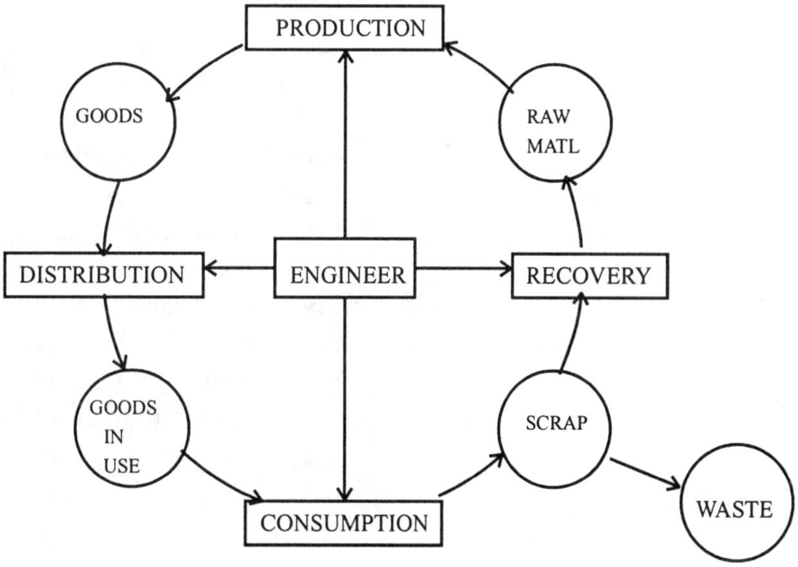

Engineers are involved in each phase of the production/consumption cycle. Raw materials must be processed. Systems must be designed for the extraction and purification of ores or the reclamation of scrap. Heavy machine design, power delivery and transportation are typical requirements at this phase. Slabs, billets, ingots, sheets, bars, beams and other metals are made into stock. Wood and wood products, cement, glass, plastics, etc. are processed. These items must be moved and warehoused while waiting for the production phase.

Finally, converting material stock into production hardware is probably the most understood aspect of engineering.

Even the consumption phase requires sales engineers to work with customers. Provisions for installation, maintenance and repair are necessary to insure expected product life.

Education requires four to six years of university level work. Emphasis is on theoretical applications of science and mathematics plus the study of standard practices within each discipline. Typical fields are excerpted from "Dictionary of Occupational Titles".

- Civil Engineer - plans, designs and directs civil engineering projects such as roads, railroads, airports, bridges, harbors, harbors, channels, dams, irrigation systems, pipelines, sewer systems and powerplants.
- Mechanical Engineer - researches, plans and designs mechanical and electro-mechanical products and systems such as instruments, controls, robots, engines, machines, and mechanical, thermal, pneumatic, hydraulic or heat transfer systems.
- Electrical Engineer - researches, develops, designs and tests electrical components, equipment and systems. Related specialties include: electric power generation, transmission and distribution, atomic power, electrical and electronic components, computer engineering and bioengineering.
- Chemical Engineer - designs equipment and develops processes for manufacturing chemicals and related products. Specialties include: heat transfer and energy conversion, petrochemicals and fuels, foods, forest products, plastics, detergents, rubber, or synthetic textiles.
- Agricultural Engineer - applies engineering technology and biological sciences to agricultural problems concerned with power and machinery, electrification, structures, soil and water conservation, harvesting, processing and transportation of agricultural products.
- Industrial Engineer - designs and implements integrated systems of personnel, materials, machinery and equipment. Specialties include: plant layout, production methods and standards, cost control, quality control, time, motion and incentive studies, and safety engineering.

Other major fields include:

- Mining and Metallurgical
- Petroleum
- Nuclear

Technicians and technologists assist the Engineer.

Technical education involves more practical levels of science and mathematics. Two years of college or community college education are required An associate degree (AAS) may be earned. Technicians work under the direction of the engineer on more detailed or practical assignments. Jobs include, inspection, repair, model building, prototype building.

Technologists usually have more college work with advanced training in a specialized area. These jobs include, drafting, specification writing, computer or controller programming, prototype testing, illustration of assembly or repair manuals, production line setup and maintenance.

Engineering team. Internal project.

A recent internal project at a large chemical plant included the following people:

- Safety Engineer
- Electrical Engineer
- Mechanical Engineer
- Structural Engineer
- Chemical Engineer
- Research Scientist

People in each field applied expertise in their own area. Design work was in a parallel mode with each engineer monitoring and approving the components as they were developed.

Consumer product design team.

Typically, a product design team is more integrated. It will include a:

- Program planner
- Market analyst
- Cost analyst
- Manufacturing specialist
- Tool designer
- Customer relations specialist
- Materials specialist
- Group of engineering designers/analysts

Team work.

All of these people must be good communicators. The team must be bonded by a clearly defined common goal. Team work is the most vital ingredient in successful product engineering. An engineer at a large factory related the following experience:

Two major design projects were undertaken at the same time. The engineering office was divided into two groups. One project was very high-profile while the other was just as important, but less glamorous. The project leader for the high-profile project chose the most dynamic and outspoken people for his leaders. The second group had very capable people who were less flamboyant.

Shortly into the initial study phase the high-profile group was beset with personality clashes and unnecessary bickering. Through all subsequent phases of the project this group was behind schedule. Initial testing turned up design flaws which led to re-design and additional expensive model testing. Eventually the project was successful but at great expense of time, money and personal health for some people.

The second group worked together extremely well. The project was as demanding as any the company had undertaken. New materials, innovative designs and forming processes were developed. The project was ahead of schedule, below budget, and expensive model testing was limited to one prototype which exceeded all tests the first try. The final product was extremely successful.

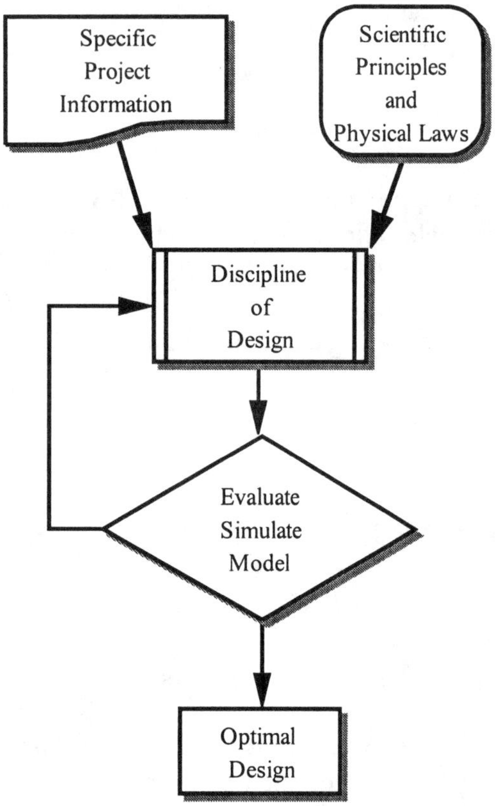

Specific project information must be obtained. Sometimes, several attempts must be made to get as clear and simple a statement of the problem as possible. Talking to users, operators, foremen, et al. may provide insight on an industrial production machine. Analyzing warranty returns and purchaser complaints will highlight weaknesses in a consumer item. Automobile companies conduct customer satisfaction surveys which are valuable feedback on a product as well as sales tools if customers are happy. A complete tear-down of a machine may be needed to determine operational problems.

In construction projects, site information dealing with soil conditions, drainage, surrounding structures, structure use and loads, etc. must be known. All levels of codes, ordinances, environmental constraints, earthquake factors and safety must be researched.

Designers must "get inside" each part of the problem in order to understand each element. *Simplify, simplify, simplify* is common advice. Sit back, visualize each detail, look at every possible interaction, explore every option, chase every fault.

How does one learn to design? Well, just like learning to swim, talk can only prepare the mental techniques and attitudes. Start by examining everything. Question why it was done that way.

Question the selection of materials. Question the power source. Question the fabrication and assembly.

"There are very few really good designers." Jim Watson. Design Engineer.

Few people have inherent ability. For most people, training in fundamentals is a start. Understanding the interrelationships of physics, chemistry, mechanics, electricity, how to simulate a process mathematically, and how to measure these quantities is essential. Understand materials.

Perception

Designers are constantly observing things. X-ray vision is essential for looking inside, back, under and through things. Filing major concepts and minute details in memory. Questioning everything. Filling notebooks with sketches, diagrams and mathematical relations. Filling the margins of books (which they own) with notes, explanations and clarifications so these bits of insight do not get lost.

Curiosity

Designers are inquisitive — tearing things apart (mentally and physically) to see how they work. Is it safe? Why does it work? How can it be made better? Why did they assemble it this way? Why is that part made that way and out of that material? Can it be repaired easily? How long will it last? Does it fail safe? How is it powered? Is it economical?

Concentration

The ability to focus on a problem. Sort out the essential elements and devalue the non-essential. Examine every possible action, reaction and possible malfunction.

Flexibility

At all points in the design process choices must be made. While common practice usually dictates a direction, options must be considered. New materials, new forming, construction or fabrication processes occur constantly. Telephones are changing shapes, becoming lighter and requiring fewer wires.

A company which manufactures precision measuring tools has been using stainless steel because it is tarnish-proof and a relatively stable material. Larger size vernier calipers are very heavy. Re-design using carbon fiber composite material provides a strong, dimensionally stable, very light, superior product.

Dissatisfaction

This is the main spur to initiate a design cycle. Some products e.g. — homes, automobiles, tools, shoes, computer keyboards — are designed right. They provide immense satisfaction in their given function. Similar items for whatever reason are frustrating to own or use. The desire to improve an item drives the design process.

Surgical instruments are carefully crafted with emphasis on shape, function and material. They are as light and strong as possible with sleek, blended forms. Balance and handling must be perfect. Any disturbing features are corrected through a tight feedback loop.

Perseverance Some inventions and designs took many years to develop. Items such as the light bulb, zipper and polaroid photography attest to the tenacious character of the inventors.

Collecting Ideas How do I get more ideas on designs? How do I know which material to use for a particular part? What is the most economical way to manufacture this part? What kind of bearing should be used — what size should it be? Should it be painted or plated? Design brings forth a never-ending series of questions.

Answers come from education, observation, research, by asking questions and, most recently, computerized *expert systems*. Handbooks and design manuals abound in every field. For example:

- Electronic Circuits Handbooks - thousands of circuits.
- "Machine Design" magazine - special issues on bearings, fasteners, plastics, materials, motors, drives, etc.
- Construction details handbooks.
- Manufacturers literature with design charts.
- Chemical formulas handbooks.
- Computer program modules.
- Patents.
- Scientific and technical journals.
- Professional association magazines and design manuals.
- Field trips.
- Government and industry standards.
- Design collections.
- Reader's Guide to Periodical Literature.
- Search Google or Lycos. Chatting on computer bulletin boards. (Post a problem and wait for all kinds of answers to come back. Some answers can be very funny.)

What the Job Seems Like ...

An engineer's view:

- 50% of the time is gathering information and promoting new ideas. This includes practical research, obtaining manufacturer's catalogs and design sheets, talking to manufacturer's representatives (who are often themselves, engineers), attending meetings and conferences, and working with models and prototypes.
- 40% of the time is justifying what you are doing. A rigorous reporting schedule must be maintained which highlights the current status of each job. Verbal reports and written materials must be prepared. Challenges must be countered. Time frames and job progress must be monitored.
- 10% of the time is real engineering. Creative design work, calculations, sketches and analysis.

"Every time there is an earthquake in California, I call to see if my bridges are still standing." Don McGlashon - retired Civil engineer.

Engineers are proud of the designs on which they worked. Every project brings new challenges, learning and rewards. Each new building, machine, process or control brought satisfaction when it was placed in use. These people like to talk about the obstacles that were faced; the creative solutions; the long hours; the teamwork, the hardships; the deadlines and the result — which often turned out better than anyone expected.

"The designer is the only person in the whole process who starts out with a clean piece of paper."
Jim Watson, Design Engineer.

Note: Don McGlashon's bridges are still standing.

Idea or Problem

Knowledge is power.

Clearly define the problem. Once the need is established, a clear description is written. This is sometimes a difficult task. Often, the problem is not clearly stated or defined. What is said or appears to be the situation is not complete. People for various reasons withhold information. Intentional withholding of information allows a person to retain a bit of control or importance. Perhaps a person simply does not state the full problem due to poor analytical or communication skills. Service advisors at automobile repair shops have to be expert interviewers to sort out the malady. Getting information requires asking questions — detailed questions, to get to the core of the problem.

Focus.

Write objectives. Stated objectives help to channel the design effort. They should be as broad as possible to avoid unnecessary restrictions on creative thinking. They may need to be rewritten as the problem becomes better defined. Contrast what is in existence with what needs to be accomplished. Write down how it can be done better. Objectives should be reviewed at each point in the design process to assure the work is staying on track. While there is a lazy tendency just to "keep it in mind", the process of writing helps the designer to clarify fuzzy thoughts.

Objectives should also list measurable outcomes. How is the design lighter, quieter, more efficient, longer lasting, safer, recyclable, better appearing, cheaper to maintain? General goals as well as directly measurable outcomes are specified. Economic analysis involving Return On Investment (ROI) may provide management with incentive to fund a project.

"You've got to get inside." Charles Muench. Chemical Engineer.

"Find a comfortable spot. Let your mind think about each element that is going on. Visualize the process. Understand each tiny detail." These comments were made during an interview of a successful designer/inventor. Many creative people use the moments before going to sleep at night to visualize a piece of a problem. The thinking process continues during sleep. Sometimes a useful idea occurs in the middle of the night. Keep a sketch pad or tape recorder handy as the thought may be gone by morning. Study the problem from every possible angle. If you are designing a bellcrank - think like a bellcrank.

Early analysis serves to verify the original problem statement. Feasibility studies look at the problem and decide if it is realistic to pursue. Estimates are prepared for sales volumes, production and distribution costs. Some type of consumer survey may be needed:

From "By Design" by Ralph Kaplan.

"With the aid of slides, he led us through the trials undertaken over a period of several months in an attempt to establish what consumers wanted in a salad dressing. The researchers had at last succeeded in isolating the qualities important to the housewife. They were ***goodness*** *and* ***creaminess****."*
"About that research", I said. "Honestly, didn't you know to begin with that people wanted goodness and creaminess in a salad dressing?"
"How could we have?" he asked. "No one knew."
"It looks obvious now, after the fact. But in today's marketplace we can't afford to guess. There's no room for intuition. We have to be sure."

Creative phase.

Several sessions may be held with the design team to discuss possible solutions or courses of action. This process may involve "brainstorming" techniques to build a list of as many solutions as possible. No matter how obscure, each idea is documented for further study. People are encouraged to be as creative as possible. Later sessions allow the team to consider ideas — a comment or fragment by another person in the group may trigger a solid solution.

Designer Eero Saarinen's approach to preliminary design.

(Paraphrased from "By Design" by Ralph Kaplan)
We looked at the program and divided it into the essential elements, which turned out to be thirty odd. And we proceeded methodically to make one hundred studies of each element. At the end of the hundred studies we tried to get the solution for that element that suited the thing best and set that up as the standard below which we would not fall in the final scheme. Then we proceeded to break down all logical combinations of those elements trying not to erode the quality we had gained in the best of the hundred single elements; and then we took those elements and began to search for logical combinations of combinations, and several of such stages before we even began to consider a plan.
.....It went on, it was a sort of brutal thing.....We reorganized all the elements, but this time with a little more experience, chose the elements a different way and proceeded ....and went right down the procedure.

At the end .... we really wept .... it looked so idiotically simple ...we thought we had blown the whole bit.

Computer programs to promote creativity

Due to the random nature of ideas and elements of a project, linking and relationships are difficult to see at times. Several new computer programs for the MACINTOSH and IBM type computers have appeared on the market. One especially useful program is INSPIRATION by Inspiration Software. This highly graphical program allows input of random thoughts in boxes. Boxes may be moved and linked in appropriate groups. At the press of a key the program presents a formal outline view. INSPIRATION is very useful during brainstorming and proposal writing sessions.

Sorting out the best.

One or several plans may survive the creative phase. More detailed design is done. Sketches, mathematical or computer simulations go deeper into the designs.

Industrial Designer. From "By Design" by Ralph Kaplan.

Don McFarland, who is trained both as an aeronautical engineer and industrial designer, describes an interesting difference in the two operating styles. An engineer, according to McFarland, works from the inside out. That is, he/she is trained to solve problems by thinking first in terms of technical details. But the industrial designer normally works from the outside in. His thinking starts with the complete product as it would be used by someone. To the engineer this may seem like starting at the end, and in a way it is. Paradoxically, this backward way of working explains why it is important for the designer to be involved in the project from the beginning ....

A popular approach to writing complex computer programs is described as "the top down, bottom up" method.

Presentation.

Findings on the viability of the project must be communicated to management. Reports with graphs, preliminary sketches, analyses and financial findings are written. A verbal report may be requested with a graphical presentation followed by a question and answer session. This group will often involve the design team, management, financial, advertising and production specialists.

Consensus.

Taking all available information into account, the "best" proposal is approved and sent to the design phase. According to Mr. Andy Lindburg at MONSANTO corporation, the group leader for the final design is often the person who primarily conceived the approved design.

Notes:

TENTATIVE DESIGNS

This phase is more expensive. Rejected designs represent losses.

One or several proposals may have shown enough promise to warrant more detailed investigation. These proposals are subjected to much more intensive design efforts. Accurate computer drawings are generated, sub-assemblies are designed and mathematical models are programmed for more detailed analysis. Work is complete enough that the chosen design can progress directly to the final design phase.

Freehand sketches from the feasibility studies are transformed to accurate drawings. Rough estimates are programmed for rigorous analysis

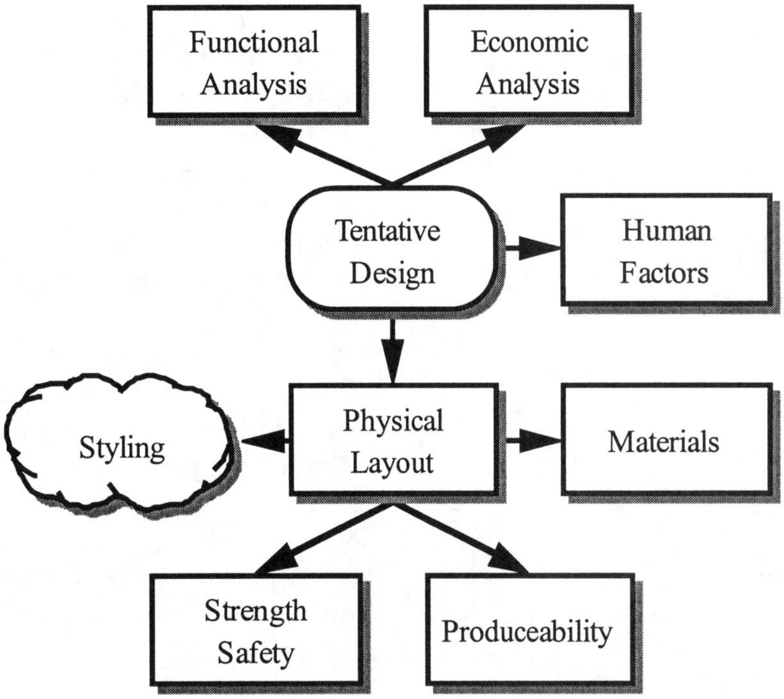

Final analysis with respect to:

- Original objectives?
- Economics
- Environmental concerns
- Performance
- Appearance/Marketability
- Safety
- Maintainability
- Producability
- Materials
- Aesthetics/Uniqueness
- Customer need

Extensive modeling and testing may be required. Scale models of chemical plants allow pipe routing and overall construction to be worked out. Scale models of airplane wings may be tested in a wind tunnel. Likewise, the dynamic performance of entire airplanes, automobiles and boats may be tested in model sizes.

Jim Watson, an aircraft designer, relates this experience:

Some of the earliest supersonic aircraft were very unstable at previously untried high speed flight conditions. One design even reached the point of tumbling through the air totally out of control. What happened? What design flaw allowed this to occur? Why did the preliminary modeling and testing not catch this problem?

Checking back, a review committee found that tentative design wind tunnel models showed stability at what was at that time considered to be adequate testing. Looking at the data, the committee noted that anomalies could be seen at the upper limits that were tried. Asked why there was no further testing, the designer replied, "We tested up to accepted points. I was worried about the top end, but we were out of time and budget. The decision was made to proceed with the existing data .... ".

Jim makes this point: **The designer must be completely sure of his work. He/she cannot allow anyone to impose ideas or constraints on the work which would compromise the intended purpose.**

There are numerous examples where designers have quit a job or a project rather than let the integrity of the design be ruined. This is not just capriciousness. It is a professional responsibility.

More reliance is being put on computer models, simulations and "expert systems". Experienced engineers have a basis for judging the results that new engineers lack. Unfortunately, almost all people tend to believe the remarkable accuracy that computers seem to possess.

Eugene Ferguson writing in "Engineering and the Minds Eye" comments:

The engineers who can "stand up" to a computer are the ones who understand that software incorporates many assumptions that cannot be easily detected by its users but which affect the validity of the results. There are a thousand points of doubt in every complex computer program. Successful computer-aided design requires the same vigilance and intuitive sense of fitness that successful designers have always depended upon when making critical design decisions.

Decision Phase

One or more designs may have passed through the preliminary phase. Each has compelling reasons to merit further development. At some point a final decision must be made as to which design goes to production.

Why did Volkswagen, Chevrolet, Pontiac, Honda and Indianapolis racers elect to put the engine at the rear of the car from time to time? What was the decision process that committed to that design? Henry Ford even built a car that steered from the rear - not the front. (Stories relate that the car was so badly behaved that it was ordered destroyed by a crew with sledge hammers.) Even today, one high performance automobile has a *slight* ability to steer the rear wheels.

Examples exist of designs that were passed over only to be revived later to become extremely successful. Witness the microprocessor. Due to late production, the original computer chips were put on a shelf with no plan for further development. Workers began to "play" with the orphaned electronics parts. Soon the wider potential of the microprocessor became apparent. A new industry was born.

Making decisions. Morris Asimow in "Introduction to Design".

"Making decisions is man's most difficult and crucial task. It lies at the heart of all human problems. It determines the direction in which human effort will be directed. If good decisions are made, man moves closer to his aspirations; if the decisions are poor, his goals recede. *Good decisions are based on the immediate evidence of one's senses, the accumulated experience of one's lifetime, the intuitive feeling for what is proper and fitting, and the recorded wisdom of civilization."*

"We assume that we have available a set of alternative plausible solutions. Each solution has associated with it sundry advantages and benefits which are expected to accrue if it is adopted. However, each solution, implying a particular course of design action, leads to various consequences or difficulties which may be more or less easy to overcome. *Thus the three elements that concern us in critical decision making, as it appears in the design process, are the alternatives, the benefits and the difficulties of implementation."*

Systematic Decision Processes.

Many plans have been proposed to aid the decision process. One of these might involve X-Y charts with a set of criteria down the side and proposals across the top. Weights varying from 1 to 10 or 1 to 100 may be assigned for each criteria and each proposal. Looking at the resulting numerical scores may point out a clear winner. More often, some intuition must be used along with the arbitrary numbers.

Options must be presented to various levels of people who

DECISION CHART

DESIGN FACTORS	WEIGHT	PROP #1	PROP #2	PROP #3
ITEM 1				
ITEM n...				
TOTALS				

ultimately approve a particular design. Sympathize with the civil engineer who must try to widen a highway through an existing town or subdivision. Which trees must be cut, which houses must go, what social effects will result from a "severe division" of an old neighborhood? How does the engineer present the proposed plans to city councils, neighborhood associations, preservation societies and environmentalists? Such decisions may be tied up for years due to advisory committees, environmental impact statements and court challenges.

Presentations include:

- Verbal presentations. Informal and formal.
- Written proposals.
- Progress reports.
- Final reports.

Formats for these reports are fairly standard.
See "Technical Reporting".

Final Design Phase

Previous design work aimed at getting and organizing information. Several approaches may may have been considered. From all ideas, tentative designs, mathematical analysis, market projections, production cost estimates, etc. one design is selected. The final design phase generates specific information on the overall product, sub-systems, assemblies and details down to the smallest parts.

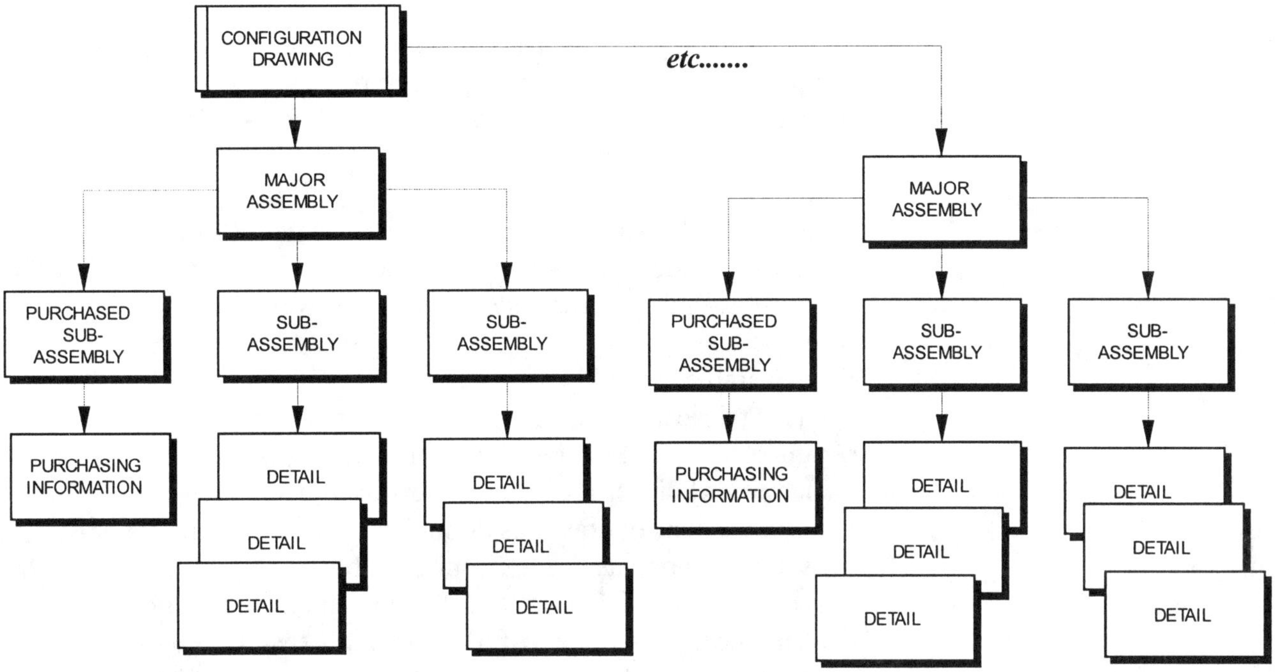

The **configuration drawing** shows overall views of the product. Each major assembly is identified. These parts may require full design or a suitable component may be purchased.

Purchased components were identified and verified during preliminary design.

Major components such as power systems, motors, drives, computers, controllers, etc. may be available from vendors. Buying these parts saves design time and production cost. Small hardware items like nuts, bolts, gears, pins, springs, are almost always purchased. A special **specification control** drawing may be generated as part of the purchase contract.

Large commitment of resources. Preliminary design required a relatively small team of engineers and scientists. Final design pulls together a much larger number of junior engineers, consultants, drafters, illustrators and technicians. The decision to go into this phase is based on extensive budget analysis, project scheduling, manpower allocation, and coordination.

Project management engineers must have considerable experience and great ability to work with people. Several very capable management computer software programs are widely used. Based on earlier **PERT** and **Critical Path Method** graphical chart planners, programs like **Microsoft Project** provide powerful management tools.

Drawings. As seen on the previous page, design works down through the major components in a tree-like structure. Sets of drawings are created following a standard format (a few are listed):

- Engineering drawing. Presents the physical and functional end product.
- Assembly drawing. Displays all parts in their assembled relationship, identifies parts (bubble callous) and parts list.
- Photo-assembly drawings may be made from models or prototypes.
- Inseparable assembly drawings depict parts that are separately fabricated, then permanently assembled (welded).
- Sub- assembly. Smaller groups of parts that must be assembled prior to insertion in the major assembly.
- Interface control drawings define the interface geometry between components. For example, the bolt sizes and locations between a wheel and tire and the axle on an automobile.
- Detail drawings. Views of parts with dimensions, notes, all manufacturing data. Monodetail drawings show one part per page. Multidetail drawings may define more than one part per page.
- Specification control drawings depict parts that are available from vendors as catalog items.
- Altered item drawing depicts a vendor supplied item that is specially modified.
- Elevation drawings show the vertical projections of buildings or the vertical profiles of equipment.
- Schematic drawings show functional connections of mechanical, electronic or chemical systems.

Detail drawings have all the dimensions needed to produce the parts. Assembly type drawings may contain only a few assembly-related dimensions.

CONFIGURATION DRAWING

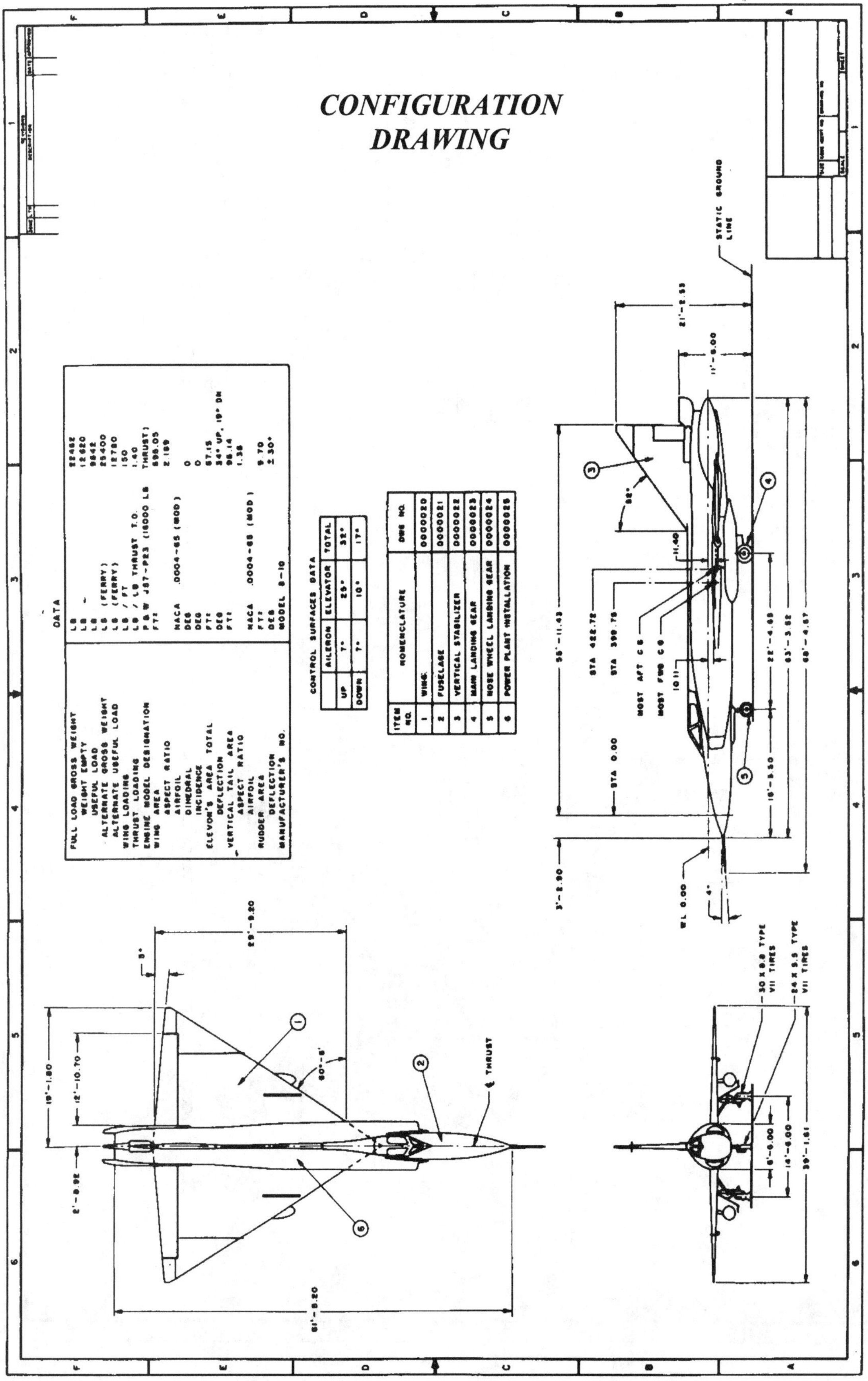

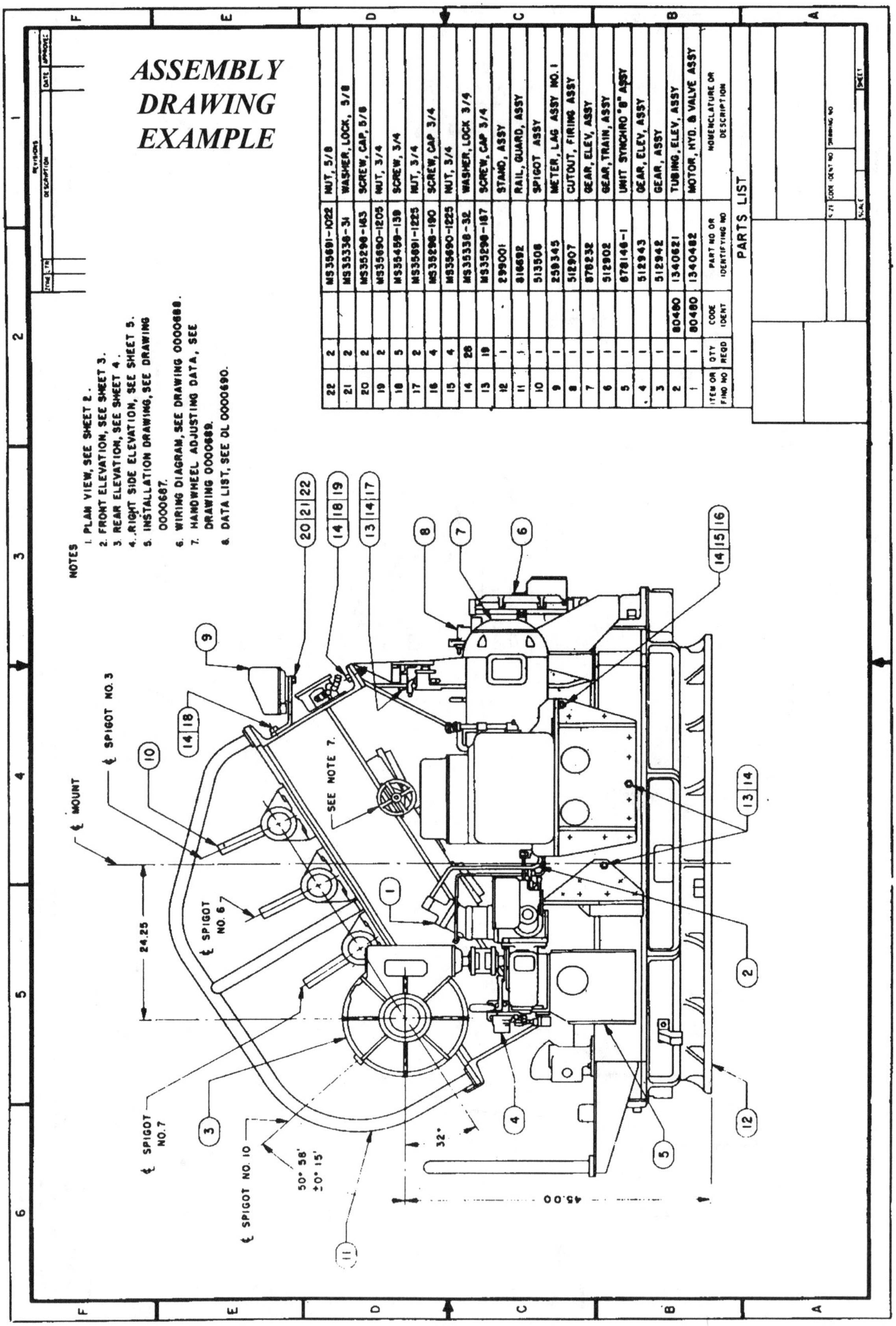
ASSEMBLY
DRAWING
EXAMPLE
NOTES
1. PLAN VIEW, SEE SHEET 2.
2. FRONT ELEVATION, SEE SHEET 3.
3. REAR ELEVATION, SEE SHEET 4.
4. RIGHT SIDE ELEVATION, SEE SHEET 5.
5. INSTALLATION DRAWING, SEE DRAWING 0000687.
6. WIRING DIAGRAM, SEE DRAWING 0000688.
7. HANDWHEEL ADJUSTING DATA, SEE DRAWING 0000689.
8. DATA LIST, SEE DL 0000690.
22 2 MS35691-1022 NUT, 5/8
21 2 MS35338-31 WASHER, LOCK, 5/8
20 2 MS35290-163 SCREW, CAP, 5/8
19 2 MS35690-1205 NUT, 3/4
18 5 MS35458-139 SCREW, 3/4
17 2 MS35691-1225 NUT, 3/4
16 4 MS35298-190 SCREW, CAP 3/4
15 4 MS35690-1225 NUT, 3/4
14 28 MS35338-32 WASHER, LOCK 3/4
13 19 MS35290-187 SCREW, CAP 3/4
12 1 299001 STAND, ASSY
11 1 816692 RAIL, GUARD, ASSY
10 1 513508 SPIGOT ASSY
9 1 299345 METER, LAG ASSY NO. 1
8 1 512907 CUTOUT, FIRING ASSY
7 1 878232 GEAR, ELEV, ASSY
6 1 512902 GEAR, TRAIN, ASSY
5 1 878146-1 UNIT SYNCHRO "B" ASSY
4 1 512943 GEAR, ELEV, ASSY
3 1 512942 GEAR, ASSY
2 1 80480 1340621 TUBING, ELEV, ASSY
1 1 80480 1340482 MOTOR, HYD, & VALVE ASSY
ITEM OR FIND NO
QTY REQD
CODE IDENT
PART NO OR IDENTIFYING NO
NOMENCLATURE OR DESCRIPTION
PARTS LIST
℄ MOUNT
℄ SPIGOT NO. 3
℄ SPIGOT NO. 6
℄ SPIGOT NO. 7
℄ SPIGOT NO. 10
SEE NOTE 7.
24.25
45.00
32°
50° 58' ±0° 15'

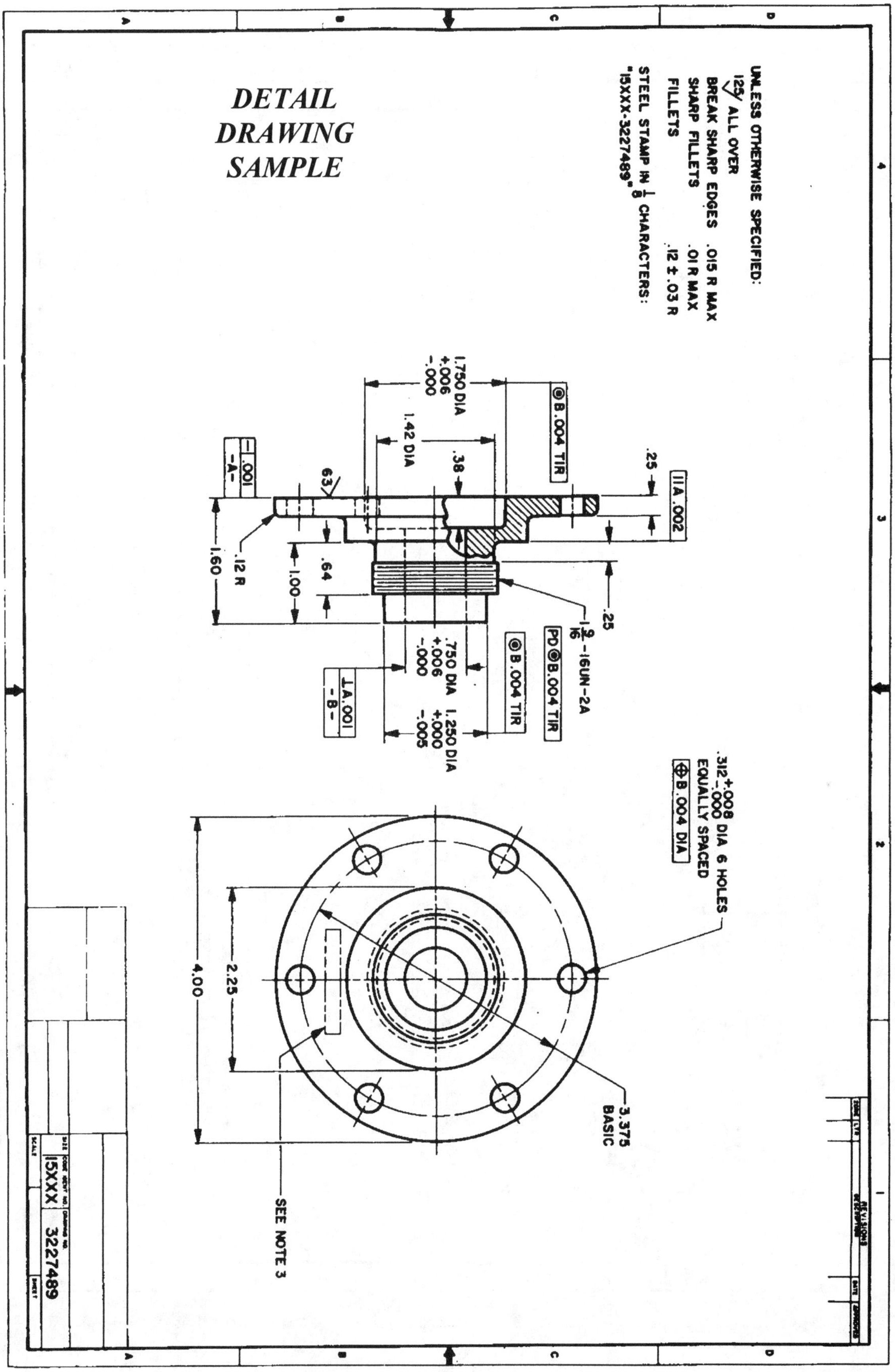
DETAIL
DRAWING
SAMPLE
UNLESS OTHERWISE SPECIFIED:
125/ ALL OVER
BREAK SHARP EDGES .015 R MAX
SHARP FILLETS .01 R MAX
FILLETS .12 ± .03 R
STEEL STAMP IN 1/8 CHARACTERS:
"15XXX-3227489"
1.750 DIA +.006 -.000
1.42 DIA
.38
.25
.002
.001
-A-
63
.12 R
1.60
1.00
.64
1 9/16 -16UN-2A
PD B .004 TIR
B .004 TIR
.750 DIA +.006 -.000
1.250 DIA +.000 -.005
A .001
-B-
.312 +.008 -.000 DIA 6 HOLES EQUALLY SPACED
B .004 DIA
4.00
2.25
3.375 BASIC
SEE NOTE 3
REVISIONS
15XXX
3227489

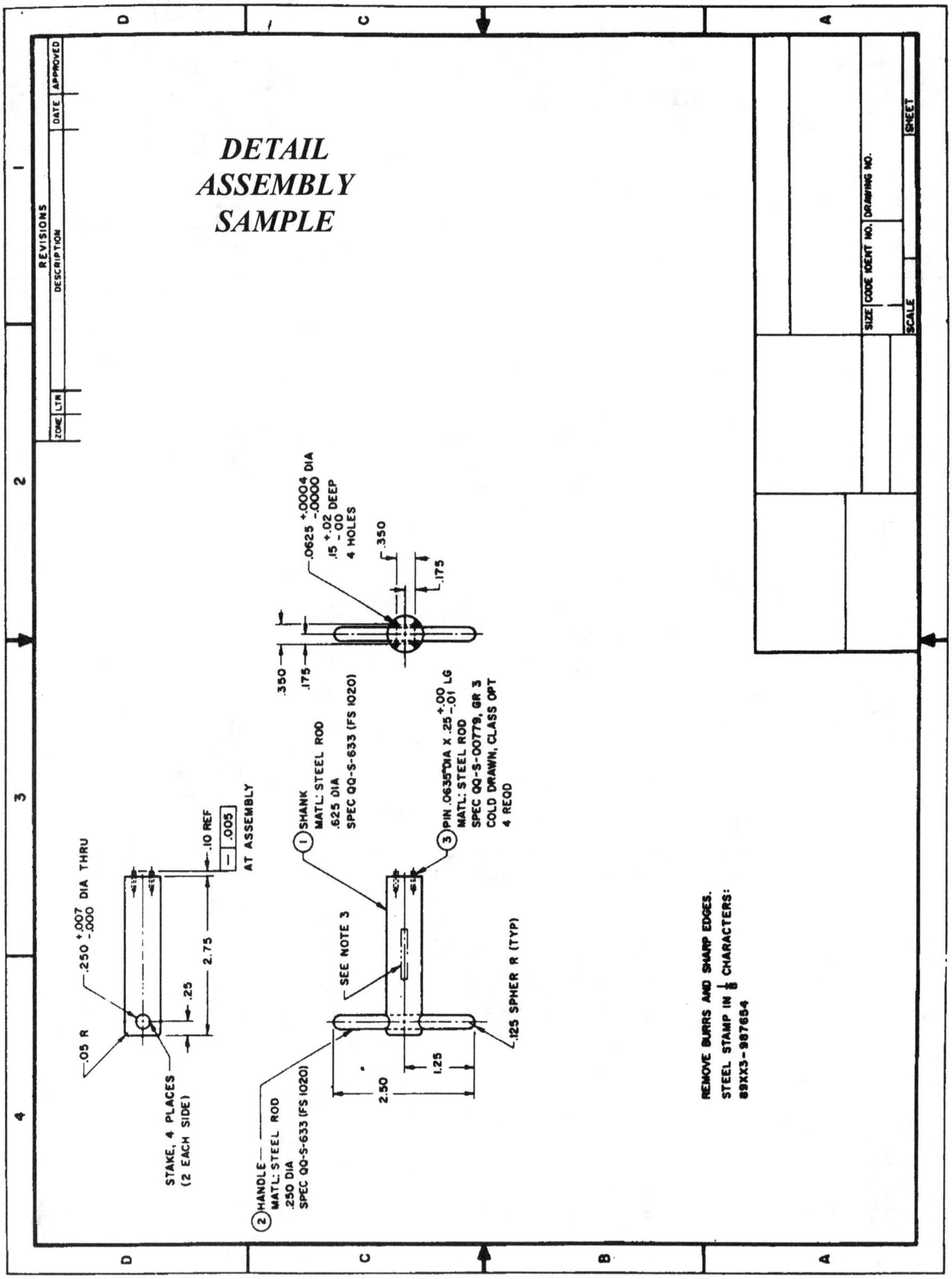
DETAIL
ASSEMBLY
SAMPLE
REVISIONS
ZONE
LTR
DESCRIPTION
DATE
APPROVED
.250 +.007 -.000 DIA THRU
.05 R
STAKE, 4 PLACES
(2 EACH SIDE)
.25
2.75
.10 REF
.005
AT ASSEMBLY
1 SHANK
MATL: STEEL ROD
.625 DIA
SPEC QQ-S-633 (FS 1020)
2 HANDLE
MATL: STEEL ROD
.250 DIA
SPEC QQ-S-633 (FS 1020)
3 PIN .0635 DIA X .25 +.00 -.01 LG
MATL: STEEL ROD
SPEC QQ-S-00779, GR 3
COLD DRAWN, CLASS OPT
4 REQD
SEE NOTE 3
.125 SPHER R (TYP)
1.25
2.50
.0625 +.0004 -.0000 DIA
.15 +.02 -.00 DEEP
4 HOLES
.350
.175
REMOVE BURRS AND SHARP EDGES.
STEEL STAMP IN 1/8 CHARACTERS:
89XX3-987654
SIZE
CODE IDENT NO.
DRAWING NO.
SCALE
SHEET

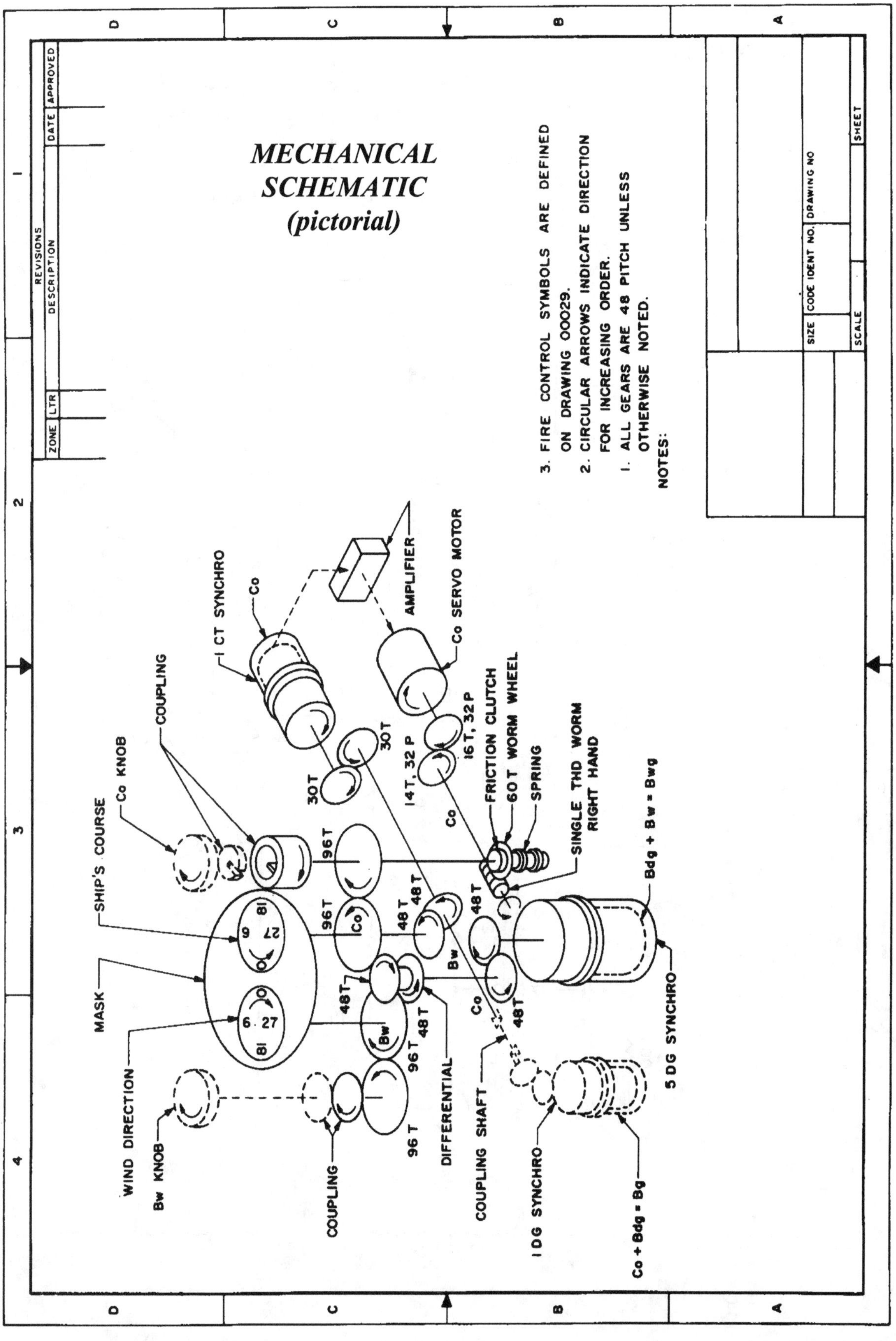
MECHANICAL
SCHEMATIC
(pictorial)
NOTES:
1. ALL GEARS ARE 48 PITCH UNLESS OTHERWISE NOTED.
2. CIRCULAR ARROWS INDICATE DIRECTION FOR INCREASING ORDER.
3. FIRE CONTROL SYMBOLS ARE DEFINED ON DRAWING 00029.
REVISIONS
ZONE
LTR
DESCRIPTION
DATE
APPROVED
SIZE
CODE IDENT NO.
DRAWING NO
SCALE
SHEET
SHIP'S COURSE
Co KNOB
COUPLING
MASK
WIND DIRECTION
Bw KNOB
1 CT SYNCHRO
Co
AMPLIFIER
Co SERVO MOTOR
30T
14T, 32 P
16T, 32 P
FRICTION CLUTCH
60T WORM WHEEL
SPRING
SINGLE THD WORM RIGHT HAND
96T
48T
96 T
DIFFERENTIAL
COUPLING SHAFT
1 DG SYNCHRO
5 DG SYNCHRO
Bdg + Bw = Bwg
Co + Bdg = Bg
Bw

electronic schematic

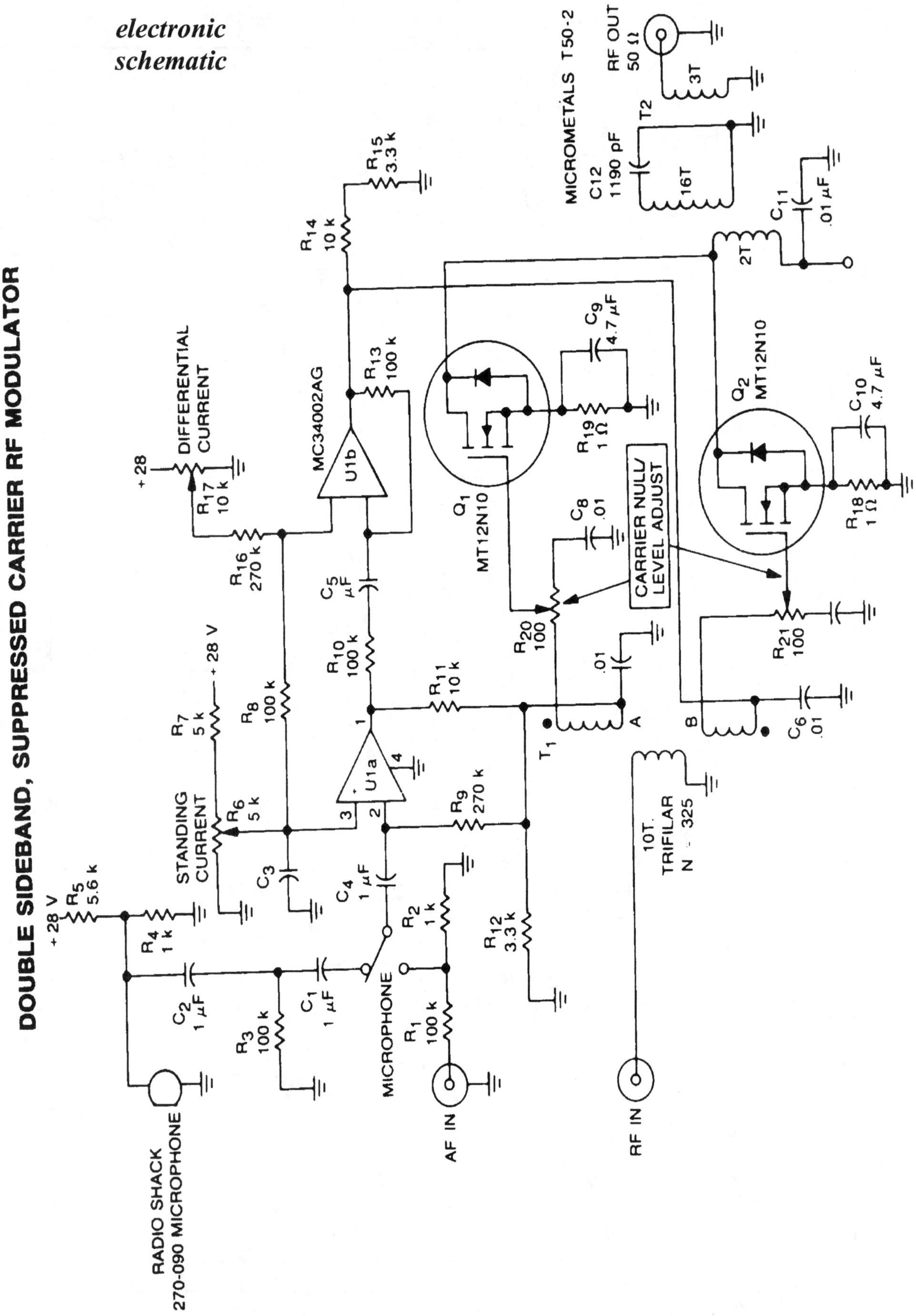

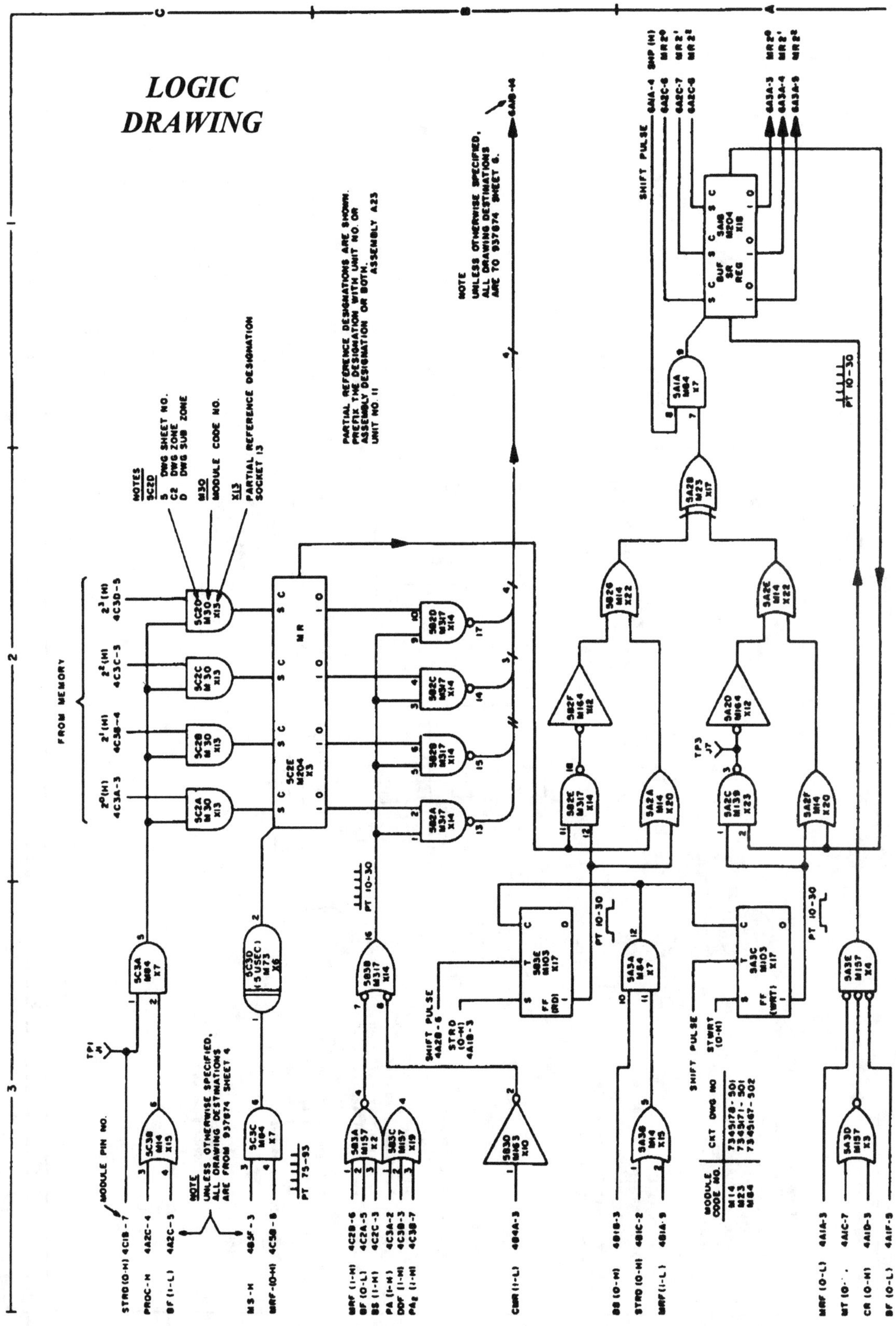
LOGIC
DRAWING

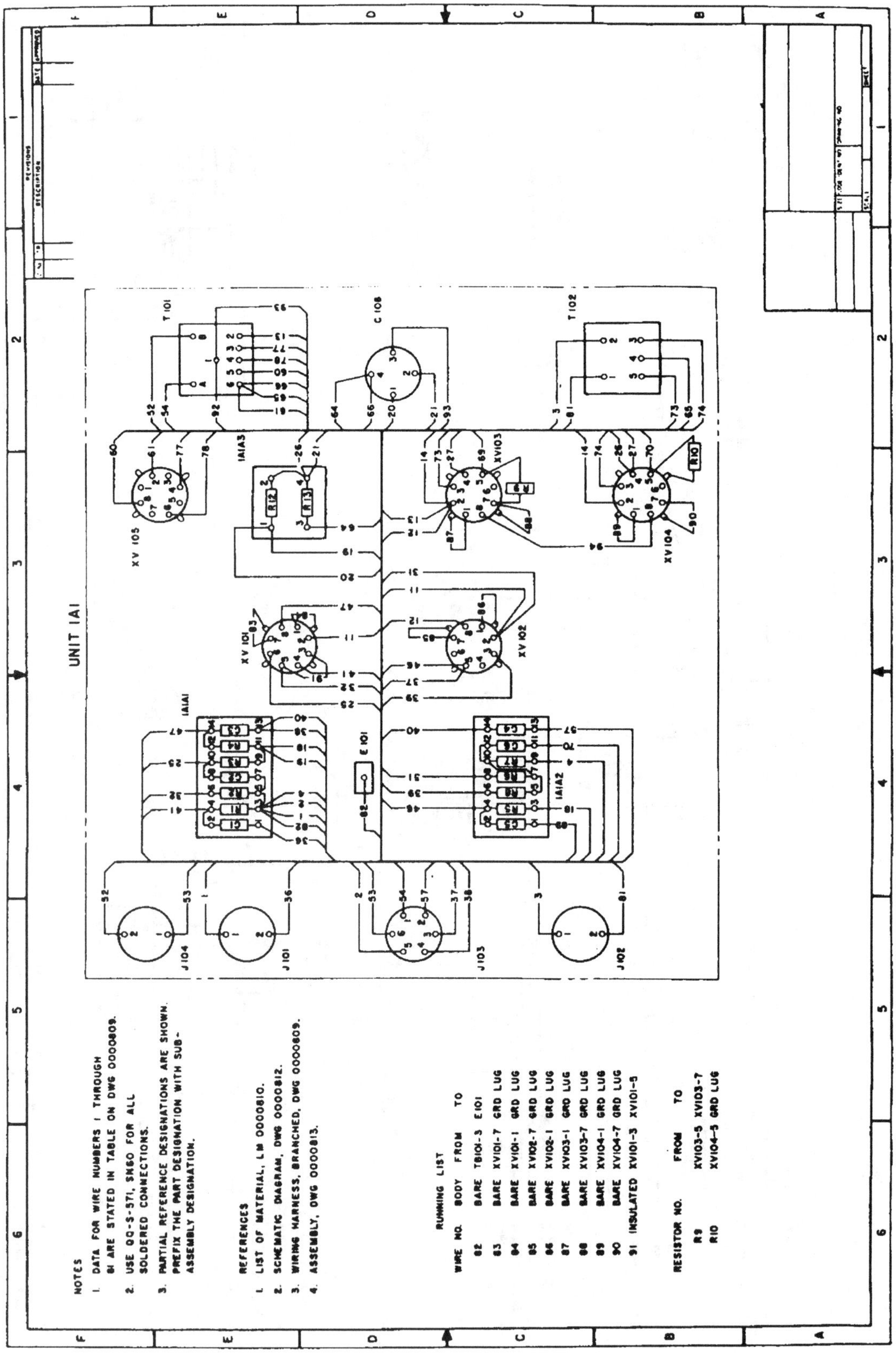
UNIT 1A1
NOTES
1. DATA FOR WIRE NUMBERS 1 THROUGH 81 ARE STATED IN TABLE ON DWG 0000809.
2. USE QQ-S-571, SN60 FOR ALL SOLDERED CONNECTIONS.
3. PARTIAL REFERENCE DESIGNATIONS ARE SHOWN. PREFIX THE PART DESIGNATION WITH SUB-ASSEMBLY DESIGNATION.
REFERENCES
1. LIST OF MATERIAL, LM 0000810.
2. SCHEMATIC DIAGRAM, DWG 0000812.
3. WIRING HARNESS, BRANCHED, DWG 0000809.
4. ASSEMBLY, DWG 0000813.
RUNNING LIST
WIRE NO. BODY FROM TO
82 BARE TB101-3 E101
83 BARE XV101-7 GRD LUG
84 BARE XV101-1 GRD LUG
85 BARE XV102-7 GRD LUG
86 BARE XV102-1 GRD LUG
87 BARE XV103-1 GRD LUG
88 BARE XV103-7 GRD LUG
89 BARE XV104-1 GRD LUG
90 BARE XV104-7 GRD LUG
91 INSULATED XV101-3 XV101-5
RESISTOR NO. FROM TO
R9 XV103-5 XV103-7
R10 XV104-5 GRD LUG
T101
C108
T102
XV105
XV101
XV102
XV103
XV104
1A1A1
1A1A2
1A1A3
E101
J104
J101
J103
J102

Electrical Connection Drawing

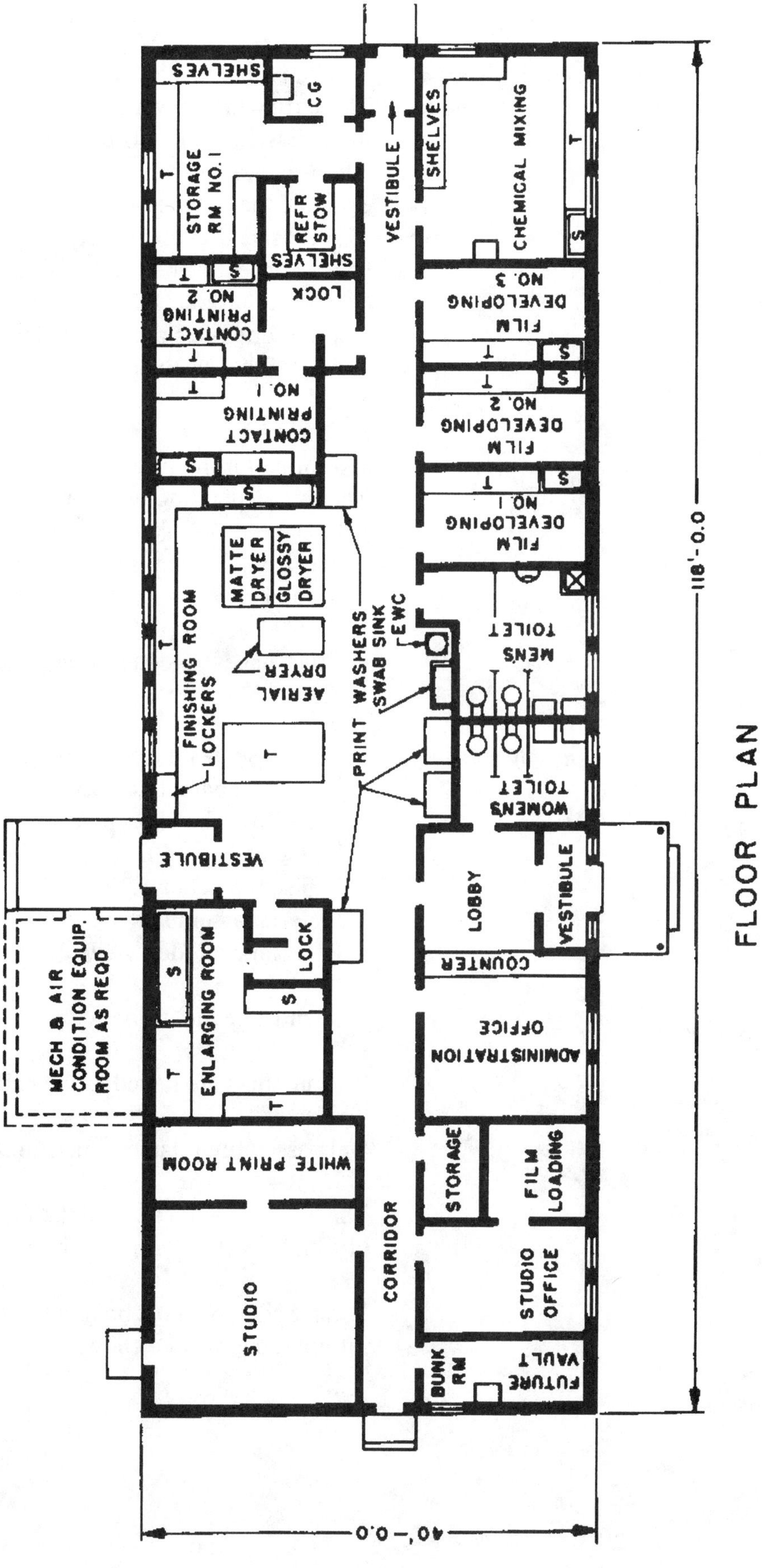

Common Details

Organization of a set of drawings.

Original drawings are kept in vaults or fireproof cabinets. These drawings never leave the design office. Some companies use a drawing numbering system which indicates the project number, part number, sheet size, etc. in coded form. For example the number 47C-2-951 might indicate product number 47; "C" size drawing sheet; sub-assembly 2 and part number 951. In order to find the drawing, the engineer would go to the cabinet for product 47 which stored "C" size drawings. All working drawings are distributed in some secondary form such as microfilms or blueprints.

Blueprints are often bound along the left side into sets of pages and rolled for mailing or carrying. The order of the pages is usually: (depending on the work)

- First page — assembly view (mechanical), block diagram (electronic), site plan (architectural).
- Second page — sub-assembly, logic diagram, elevation views.
- Other pages — detail drawings, schematic diagrams, floor plans.

Assembly drawing.

This page shows the entire product as it would actually fit together. All parts must be seen, so one of several types of drawings may be chosen:

- Single orthographic (front) view if all parts are visible.
- Full section cutaway view.
- Pictorial assembled view.
- Pictorial exploded view.

Parts must be identified with a "bubble callous.

Part numbers are keyed to the parts list.

Only assembly related dimensions and notes are shown.

The parts list is located on the lower left corner of the title block. Parts start with 1 at the bottom and go up.

Purchased parts must be identified and all purchasing information must be shown.

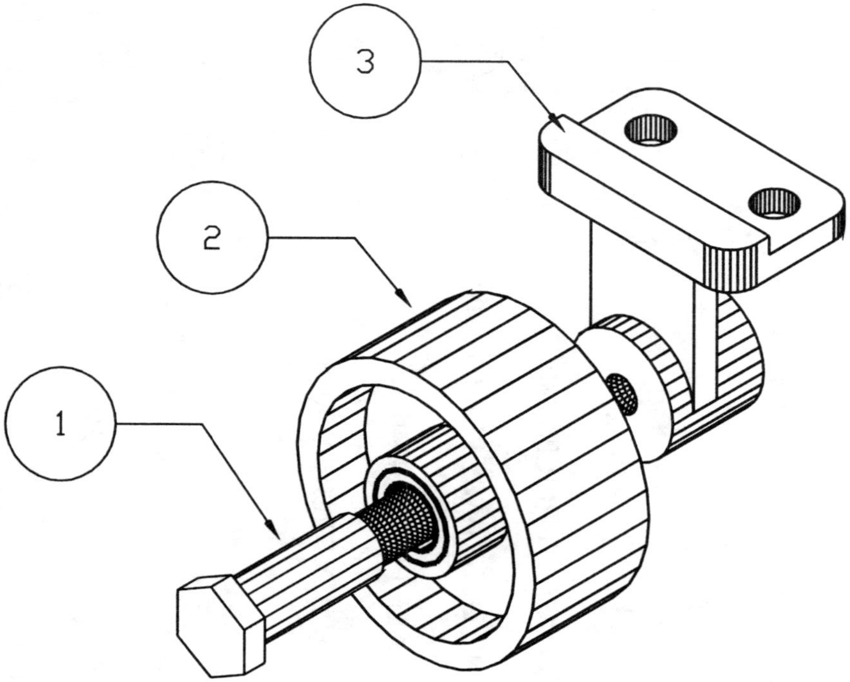

3. APPLY MDS GREASE TO BEARING SURFACES.
2. APPLY LOCTITE TO THREADS ON ASSEMBLY
1. DEGREASE THREADS PER PS9014
NOTES:

ITEM	PART NAME	REQD	MATL
3	BASE SUPPORT	1	CAST IRON
2	WHEEL ASSBY	1	
1	SHOULDER BOLT	1	SAE 1116

COMPANY / SCHOOL NAME	DRAWN BY: LAST, F	DATE:	SHEET 1 OF

Detail drawings. Each part which must be fabricated or modified must be drawn. Many companies prefer that only one part be placed on each (smaller) page. However, if the procedure is to use all the same (larger) size pages in a drawing set, multiple parts may be placed on the same sheet.

- Parts are identified with the same bubble callout number from the assembly page.
- Material is specified.
- Number of parts per assembly is specified.
- Special manufacturing notes are included with (1) at the bottom.
- The detail drawing must stand alone. All information to manufacture or modify the part must be included.
- Views must be in correct location and projection. Enough views must be included to fully describe the shape of the part in visible outlines. Partial views, auxiliary views and sectional views may be needed where complex shapes are not seen in the regular views.

Larger companies may have special Process Specifications which describe exact methods of machining, fabrication, cleaning, welding, etc. A detail drawing note may reference a process specification rather than recording a long series of instructions:

BREAK ALL SHARP EDGES PER PS 1000

INSPECT PER PS 2012

ZYGLO TEST PER PS 2700

MAGNAFLUX TEST PER PS 2800

CLEAN AND DEGREASE PER PS 3000

NICKEL PLATE PER PS 4040

PRIME AND PAINT PER PS 5000

Typical Student Project

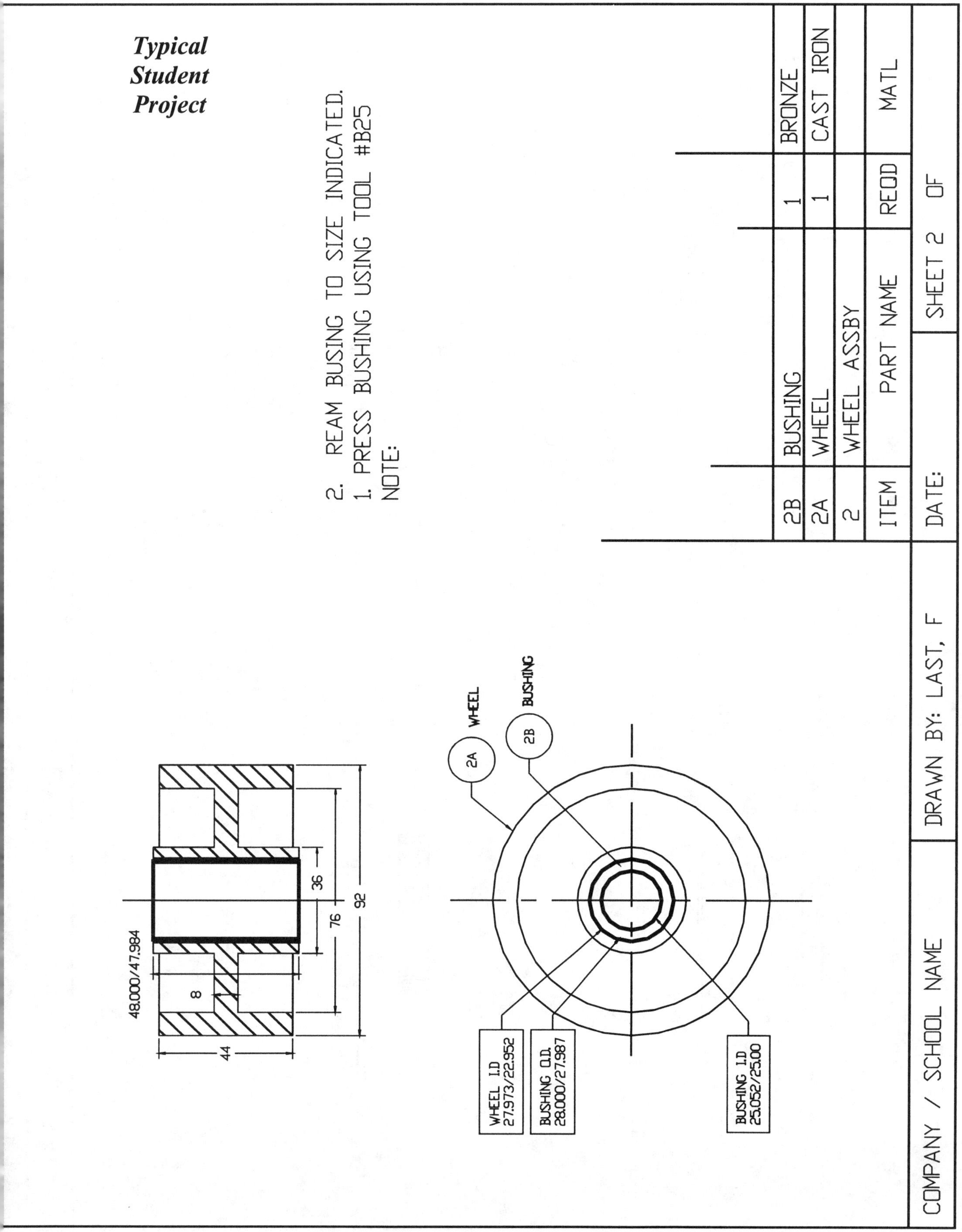

Typical Student Project

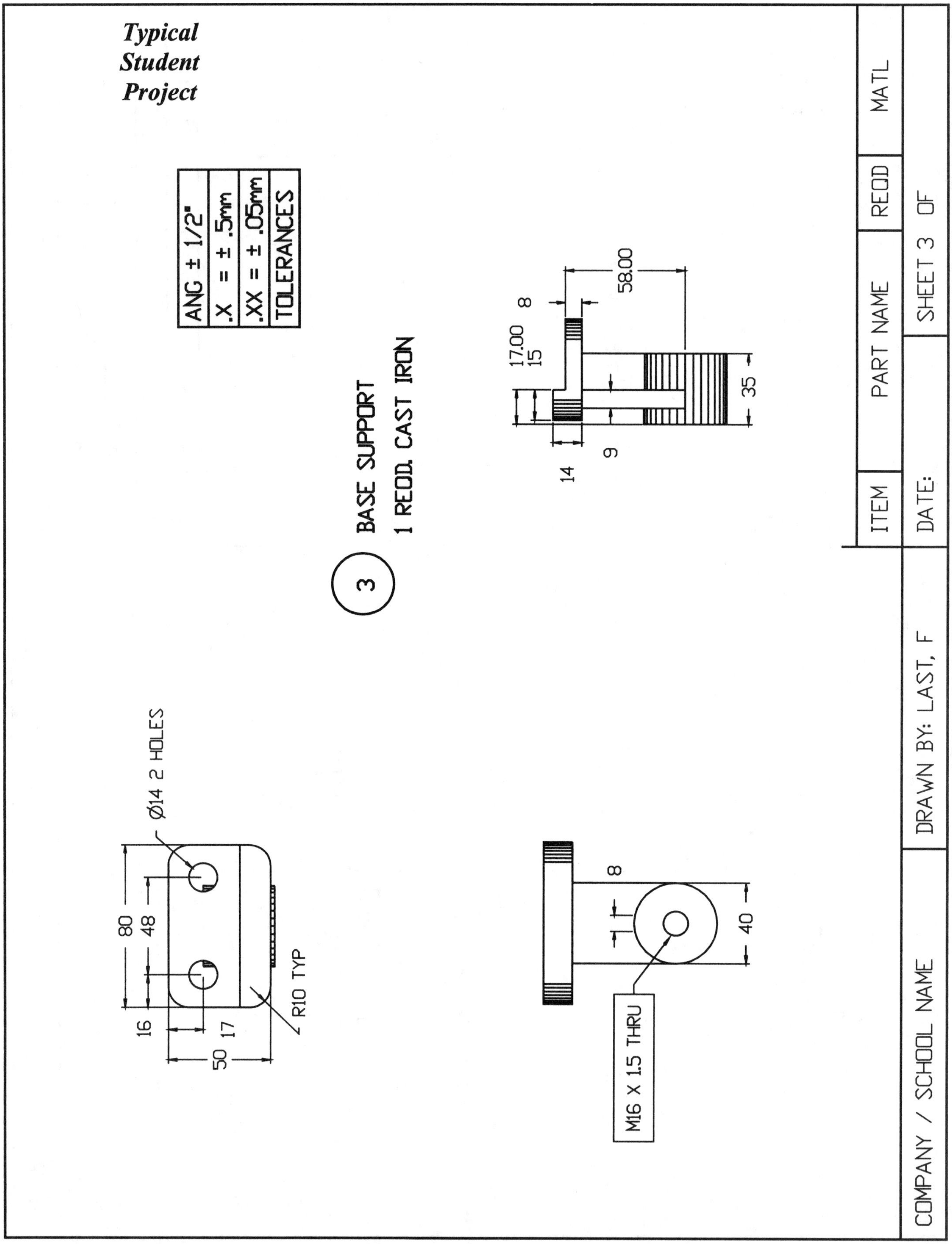

Typical Student Project

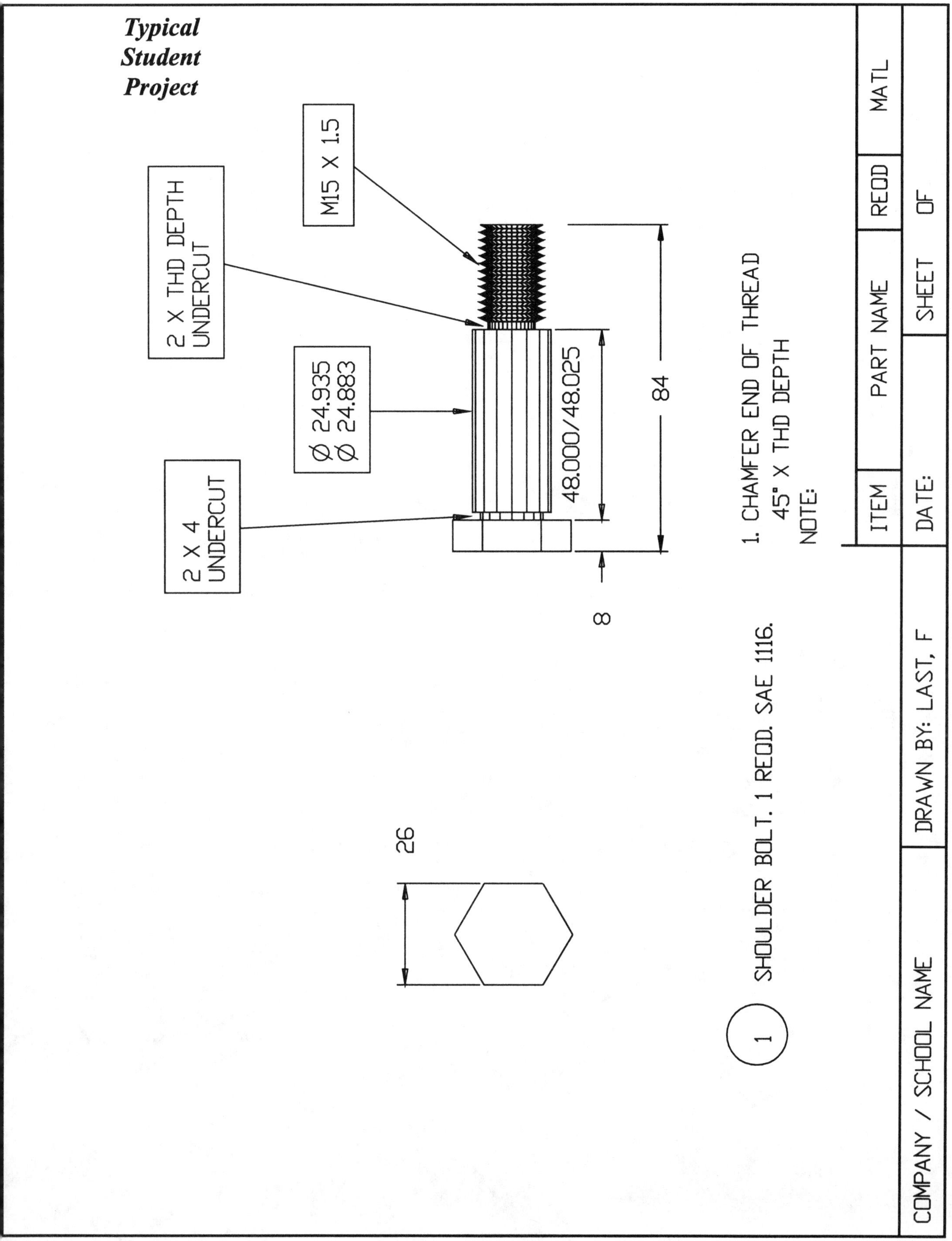

Engineering Project Management

First day on the job ...

What is it going to be like when I start a new job? What will I do? How do I get started on the right foot? These questions have gone through the mind of every person reporting for a new job. Engineering jobs require a high level of education, so a college degree or substantial course work is the qualification. But, how does one learn the particulars? How do you get familiar with the current work or project? Often, newly hired engineers spend the first day filling out forms, taking physicals, being briefed on the health plans and other benefits. Get fingerprinted and photographed for a badge. A high company official may drop by to offer congratulations and a brief pep-talk. Tools, office supplies, and other items may be provided in a box. Tours and overwhelming amounts of information finish the day.

Project assignment...

Second day. Show up at the project group area, box of supplies in hand, a little late because the place is big and the directions were minimal. Introduce yourself to the person at the first desk. A few minutes of confusion and phone calls follow. Somehow, your presence was not exactly expected at this time. You are welcome but, they have to move some desks to get yours in place, hopefully, by lunch time. Someone hands you a technical journal to read. Specific stuff. Nothing looks familiar. The project manager bustles in, says a brief hello and in a few minutes your group leader pops in. More tours. Meet a lot of people. Look at prototypes. Discuss major design objectives, current progress and problems.After noon, a colleague sits down and talks you through all the design manuals, handbooks, process specification manuals, phone books, design group hierarchy, auxiliary services, etc. Information overload, again. Everyone is very busy. Information flows in a tightly organized pattern.

First task...

Drawing changes. The best way to get involved in a project is to work with the drawings. The overall design may be studied from the configuration drawing on down. Individual components and their relationships can be seen. Changes are required as the design progresses. Clearances, access for fabrication or service may be needed. A small change in one area may precipitate a number of small adjustments down the line. New people can handle these important tasks while becoming more familiar with the entire system. All design changes are carefully monitored by the group leader.

A new person should study each assignment very carefully to be sure the full design intent is understood. Design changes, no matter how small, should conform the the existing "design style". The changes must look like they belong there.

- Learn the parts of the design.
- Why is the change needed?
- What is the process of getting it done?
  Engineering orders, drawing change notices, etc.
- What are the local design practices and preferences?
- Simple re-design may cause strength or fabrication problems.
- This is a good way for management to observe the performance, creativity and work ethic of a new person.
- The new person can observe the people in the group, find knowledgeable individuals and find a mentor.

Design suggestions may be sketched or input to a computer in the form of a temporary overlay. The lead engineer will observe and approve the design changes before they are coded in final form.

Accepting criticism ...

In most cases the first attempt will not be acceptable. The new engineer must be able to take valid suggestions. Often, the re-design is based on too little information or too hasty a solution. ***Some lead engineers can be very blunt when the design does not suit.*** This in itself may be a way of weeding out those who cannot do the job or those who do not fully research the problem.

Project Scheduling

How does a project schedule evolve? As a new person on a job, the overall timing and progress monitoring may not be clear. How does management know how much time to allocate? What happens if the work is not completed on schedule?

Project definition...

Preliminary designs by a small group of engineers and scientists will have been approved by management. Those designs will have formulated the approximate needs of the major assemblies. Calculations on power, loads, heat, strength, performance, et. al. will have been approved. A master configuration drawing will have been produced which specifies the major assemblies.

"Bottom up" estimates ...

Lead Engineers are assigned to the major assemblies. (In an automobile there might be power, chassis, body and interior major assemblies.) Those engineers might further sub-divide the work. Each senior engineer must ***visualize the design*** needed for each part of the assembly and estimate the time required for design, analysis, testing and re-design. Starting at the smallest items and working to the top, estimates are summed. The final estimate must represent a fair and honest appraisal of the work needed. Engineers are often too optimistic and management may double the figures for safety. Good estimators with considerable experience can be very accurate at this task.

Planning calendars are created.
- Drawing layout completion date.
- Submittal of design date.
- Revision date.
- Final signoff date.

Critical timing dictates manpower.

Most projects must be done in the shortest possible time to meet customer demands or competition. The time required for a job is roughly man-hours divided by people. Due to the tree-like structure of many design projects, some activities must be completed before others can begin. You cannot put the floor in a building until the foundation is complete.

Multiple tasks ....

Engineers are given assignments for various portions of the design based on their experience and training. Time for completion is set. The engineer may be working on several jobs at one time. Problems in meeting deadlines occur because:
- Many approvals must be met.
- Re-design needed because the original was less than optimum.
- Designers "fall in love" with their ideas or work. They are reticent to look at alternatives or accept changes.
- It is too easy to overlook details.
- Preliminary work from another design group is behind schedule.

Getting the job done on time...

This may require extra manpower, extra long hours or whatever it takes to meet deadlines. It is not uncommon to work seven days a week or extra hours each day to overcome unexpected delays.

Extra days beyond a deadline is not an option!

Monitoring Progress

Most projects have at least a weekly meeting to discuss progress and difficulties. Each lead engineer must discuss the status of the work in his group. Daily meetings may be needed near the end of a project as the timing becomes critical. "I had a thirty minute stand-up meeting every morning at 7am", commented one engineer. "We were very close on time and this allowed us to talk briefly before the day got started."

"Walk Around" management technique.

Managers often allocate some time to walk through design groups, talking to lead engineers and designers. Problems may be discovered before too much time is spent. Questions and answer sessions provide quick insight and information. Expect probing and challenging questions if you work for a sharp lead engineer. Managers and designers may visit construction sites or fabrication lines to look at the product being built.

Design Approval

Committee approval...

Each part of the design must be approved before fabrication can begin. Each engineer must submit his/her design and defend it against one or more management people. Verbal and graphical skills are very important.

- Designer must SELL the design to the boss.
- Communication skills are critical. Thorough preparation is necessary. Anticipate challenges.
- Designer must earn the respect of management. Stand for what is right. Accept responsibility when appropriate.
- Designer must prove his/her competence. Selling is easier thereafter.
- Designer must be completely honest.
- Designer must take charge. A committee may challenge a good design just to test the designer. Or, a committee may challenge a design for the wrong reasons based on too little information.
- Designer must resist changes if not for a good reason.
- Designer must get along with other people. There is no need to be disagreeable. The adversary atmosphere has a purpose.

Jim Watson, designer, comments:

"There is no room for prima donnas."
"There is no room for deception."

Technical Reporting

An overview of the major types of communication is shown. Technical communication involves combinations of verbal, visual and written methods. Choosing the most appropriate means of presenting information is an important task. As stated earlier, one engineer saw 40% of his time as justifying what he is doing. Clear presentation of work is extremely important.

Written Communications

Memos and Sketches. Written communication is essential when exacting information must be transmitted. Hand lettered memos and freehand sketches provide clear, readable and archiveable proof that information was transmitted or received. Many times these fragments of a design process become very important. Patent applications, lawsuits and arguments may be settled based of bits and pieces documented by memos. Verbal orders are not enough. "Write it down" is the rule.

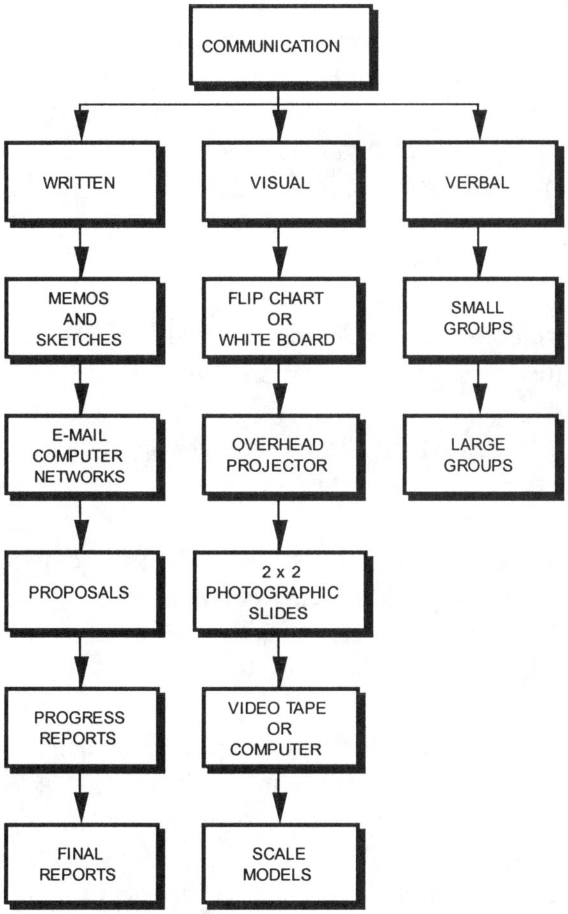

E-MAIL. Computer networks provide the opportunity to exchange information very quickly. Many design groups are organized with a computer workstation for each person and a common network connection. Conventional memos and sketches may take a day or more to reach the recipient. E-MAIL can be sent to and read by the recipient almost instantly. Replies may be forwarded back to the sender very quickly. Drawings, scanned images or other graphics may be attached to an E-MAIL message. Or, the recipient may be able to log directly in to the document or drawing for a first-hand view.

Computer connections now span the earth. Design groups in many parts of the world have direct connection to the central documents and drawings. Visual teleconferences may be set up when direct person to person discussion is needed. Design meetings which formerly required time and expensive travel are now possible within minutes worldwide - if needed.

Proposals. Formal application must be made to document the need for a design project and to request initial funding.

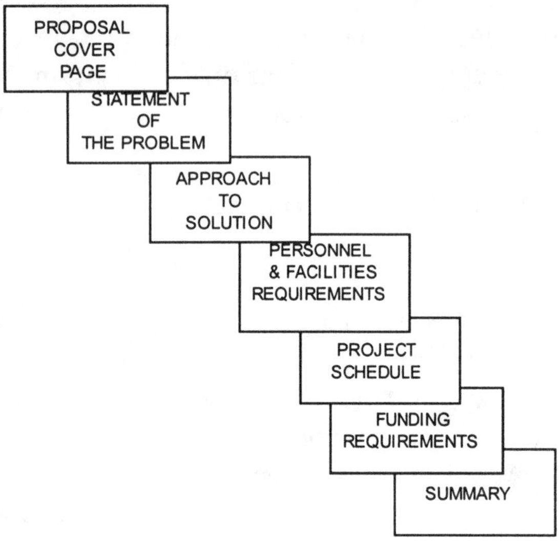

Cover Page. Include a title for the project, the division, department or other identifying information, the names of the applicants and the date submitted.

Write a clear statement of the problem. Define the need. Identify compelling reasons why the project should be undertaken.

Document the approach to be used. Identify the alternatives and how each will be considered. Define the criteria for decision making.

Make projections for the numbers and types of workers needed. Identify laboratory, research, model building needs. List the offices, office equipment, communications requirements. Identify overhead expenses.

Put together a project schedule. Work out deadlines. (Several excellent computer programs are available for this task.)

Work out a budget. (Engineers tend to be very optimistic here. Most comptrollers project larger budgets.)

Prepare the proposal in a very professional manner. Remember the audience — owners, stockholders, technical and non-technical people, bureaucrats .... Keep it simple but technically complete.

Progress Reports. Some type of reporting schedule is usually imposed. One company has a weekly meeting early Friday morning. Verbal and short written reports are presented. Critical problems are discussed. Focus for the next week's work is outlined. An open discussion period allows time for one-on-one or small group meetings. The company provides healthy food, juices and bran muffins — coffee and donuts are prohibited.

Final Report. This document is a complete description of the project. It often requires a significant amount of time and effort to assemble.

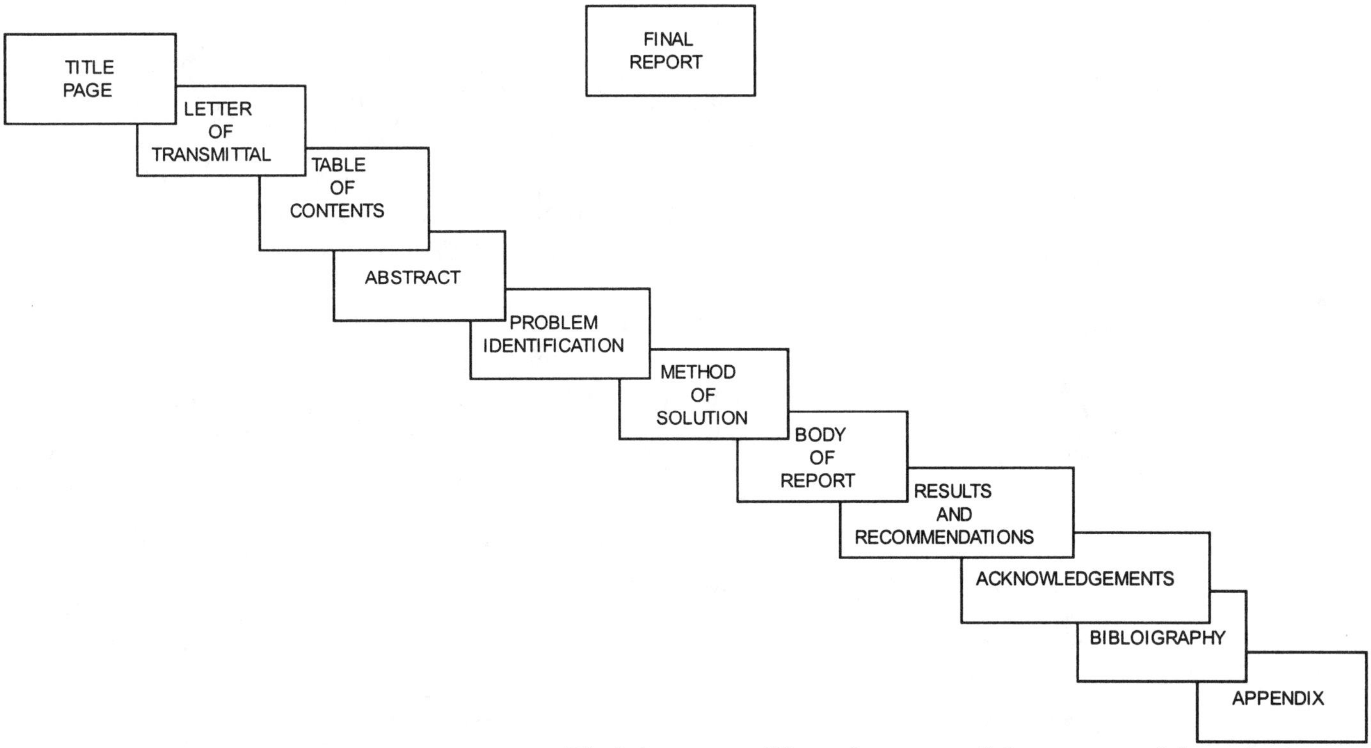

- Bind the report. Place the name of the report and the principal investigators on the cover.
- The letter of transmittal is a formal letter containing a brief statement of the problem.
- The title page includes the name of the project and a more complete list of the team members.
- Table of Contents and Table of Illustrations.
- Problem Identification
- Method of Solution.
- The body of the report is a rigorous description of the design process, the data collected, research conducted, decisions and steps to solution.
- Results and recommendations.
- Acknowledgments to persons and companies who aided in an extended manner.
- Bibliography.
- Appendix.

Lab reports, drawings, analysis, etc. are contained in the appendix.

Visual Communication

Use a visual communication method which is appropriate for the group to be assembled.

If the group is small and informal, use a flip chart or whiteboard. These items allow people in a smaller group to see clearly. Each person may wish to get up and sketch an idea or make suggestions. Use colored pens to block individual items. Use a contrasting color to indicate changes or corrections.

XEROX corporation manufacturers a special white board which attaches to a printer. Sketches on the board may be directly captured on printed pages. Multiple copies may be requested for each person in the group.

Larger groups may require and overhead projector, 2 x 2 slide projector or video projection system.

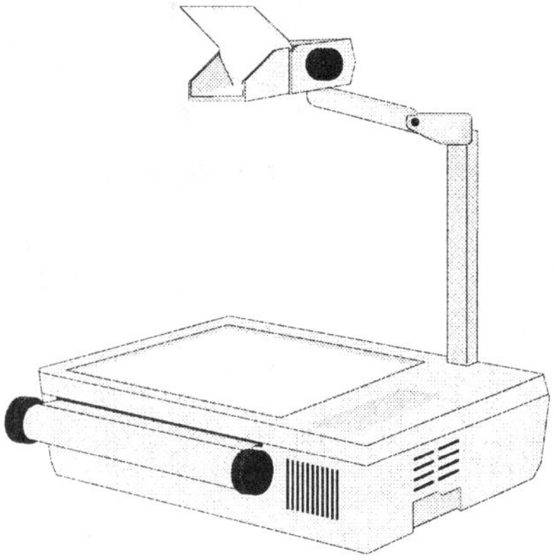

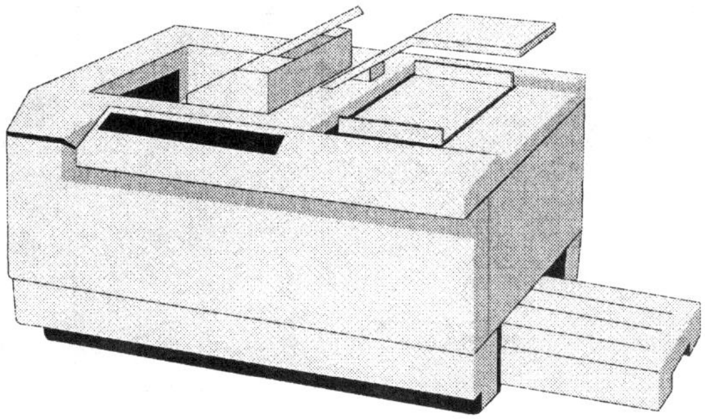

Overhead slides may be created using a Xerox type copier, laser printer, color laser printer, or inkjet type printer. Slides are relatively easy to create using drawing plots or computer illustration programs.

Keep slides open and simple. Too much information or too many lines make slides difficult to see by the audience. Test your slides by printing a test. Go into the room where the presentation is to be given. Set the light level.

- Check the line thickness. Are lines visible?
- Colors. Are colors visible - too dark - too light?
- Text height. Is text readable from the back of the room?

2 x 2 slides are useful where objects exist. Pictures may be taken using a wide variety of lenses from close up to telephoto. If lines or text are to be photographed, be sure the thickness is adequate on the original artwork. Text and line thickness on 2 x 2 slides must be tested for proper viewing. As with any program, be sure slides are in the correct order and the correct orientation!

Video tape programs may be produced at low cost. While there is a large range of quality depending on the experience of the producers, valuable technical information may be captured. Video editors and title electronics systems are available for standard computers at relatively low cost. These devices control editing and allow input of text when producing the final tape copy.

Computer graphics may be captured from CAD programs using the **slide** command. This command will save a copy of the current picture on the screen. These images are usually small graphics files. Many of these may be stored on a single floppy disk. First zoom in to the area to be shown or organize the screen image. Then, capture a slide. This is an excellent way to document progress on a project or to show alternative design ideas. Slide files may be sent through the mail or they may be transmitted via modem to remote locations.

Computer slide programs may be played back without the need for the CAD software. Also, slides may often be imported into word processors for report writing.

An LCD projector or LCD tablet on an overhead projector may be connected to a computer for large audience viewing. (Liquid Crystal Display).

Scale models are often built in order to clearly see the proportions and shapes from many angles. Architects build models for large projects - churches, athletic complexes, renovations, etc. Chemical plants are extremely complex systems of pipes, tanks and control systems. Models help to visualize pipe connections.

Regardless of the type of meeting, a careful plan of the presentation is needed. Start with an outline. Write key thoughts or points. For larger meetings, make a set of 3 x 5 cards with key words or thoughts. Number the cards so you are sure of the order.

Large meetings often require an agenda or outline to focus on the topic. The agenda may be sent to each participant ahead of time so everyone is prepared. This is an important means of keeping the meeting on track

If you are nervous, try to imagine the people around you in a humorous manner. Dress nicely in comfortable clothes. If the presentation is really important, put together your best "power" outfit. Carry a good-luck token — it couldn't hurt.

Concurrent Engineering

Projects used to pass through a series of design groups ...

Traditional engineering projects involved three distinct phases: concept phase, design phase and production phase. These activities progressed in sequence through a number of people. Some engineers joke about the ***over the wall*** progression of engineering information — "we would approve the stress calculations and pass the drawings on to the people in the next office". This serial flow of effort tends to throttle a project. Long design/production cycles occur.

- Too few people are aware of the full design objectives.
- There is too little opportunity for feedback and too little opportunity to go back to early design for improvement.
- Production is slowed because critical parts may require special materials, machining or tooling. Some tools may require many months delay before delivery from vendors.

CAD/CAM, CIM, CE, TQM, JIT, FMS, MRP ....are strategies for optimizing the design/build process.

Computer assisted design creates a very accurate numerical part description database. This information was not previously available. The Concurrent Engineering concept allows all people and workgroups to access data early in the design process. Close monitoring of unusual or critical features allows lead time for solving production problems near the end of the process. Some estimates site a reduction of 50% or more in the design/production time required with higher product quality.

- CAD/CAM Computer Aided Design - Computer Aided Machining.
- CIM Computer Integrated Manufacturing.
- CE Concurrent Engineering.
- TQM Total Quality Management.
- JIT Just In Time.
- FMS Flexible Manufacturing Systems.
- MRP Material Resources Planning.

3D Modeling vs 2D drawing

Engineering drawings in the form of orthographic projections have been the main design documentation since the Industrial Revolution. Engineers created two dimensional drawings, auxiliary views, section views and dimensions to describe design intent. At one time these drawings were very ornate - they were laboriously inked on cloth, vellum or mylar film. Engineers learned geometric construction, hand lettering and inking techniques because they spent many hours at the drawing board. Drawing changes were time-consuming and costly. Drawings must be checked and released by senior engineers for other steps to occur.

When Solid Modeling is part of an integrated system, it becomes the basis for Concurrent Engineering.

Engineers recognize that they can work more effectively if they change the traditional design progression. New tools are available. By creating master designs as 3D computer solid models, other design groups can access the data, monitor design progress and provide input for critical features. Product performance, appearance, cost, efficiency, safety and ecology can be improved in a timely manner.

"We even included our customers in the design process. They said appearance, weight and size were very important. We never imagined that appearance for a part buried way down with other parts was of that much concern", said one company owner.

Master Model.

Engineering groups can work more efficiently if there is an accessable master model. Early in the design process, a 3D solid model is created. Careful control of changes to the model involve permissions granted during drawing login. Certain key people need to be able to change the model as the design progresses. Some system for notifying all users must be in place.

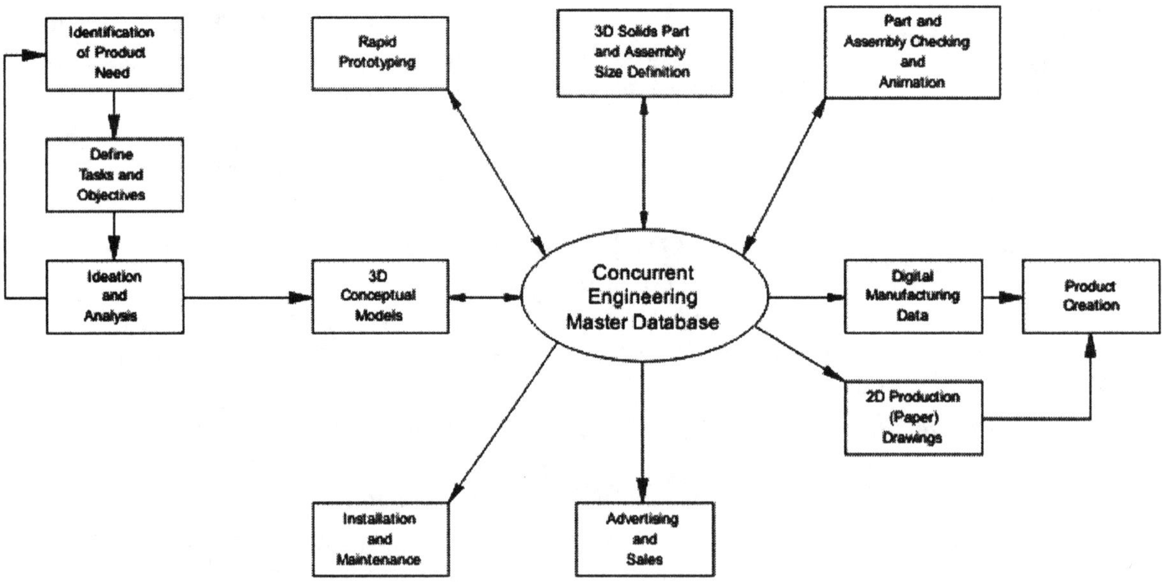

Associativity keeps everyone up to date.

Drawings in use by various members of the design group must be constantly updated. Computer networks allow many persons to access or use the master drawing. By invoking file locking, only one person at a time may make changes. But, all persons logged in to the drawing can see the changes as they occur.

Major changes may trigger many recalculations by strength, safety or production groups. The re-verification process may occur over much shorter time than was ever before possible.

Controlling changes.

Two methods of associativity have been implemented.

- Unidirectional. Master drawings may be viewed by workgroups. These groups may add their own information locally in the form of red-lines and markups but they may not alter the master. All numerical and graphical data is accessible. Changes in the master are reflected the next time the master is updated.
- Bidirectional. Master drawings may be changed by anyone with the proper access. The master drawing is open directly on line. Access permissions are granted by the senior designers.

Concurrent Engineering is an essential process for the future. Competition from traditional directions as well as foreign countries makes better, more efficient designs, at lower cost, at shorter times, a high priority for economic survival.

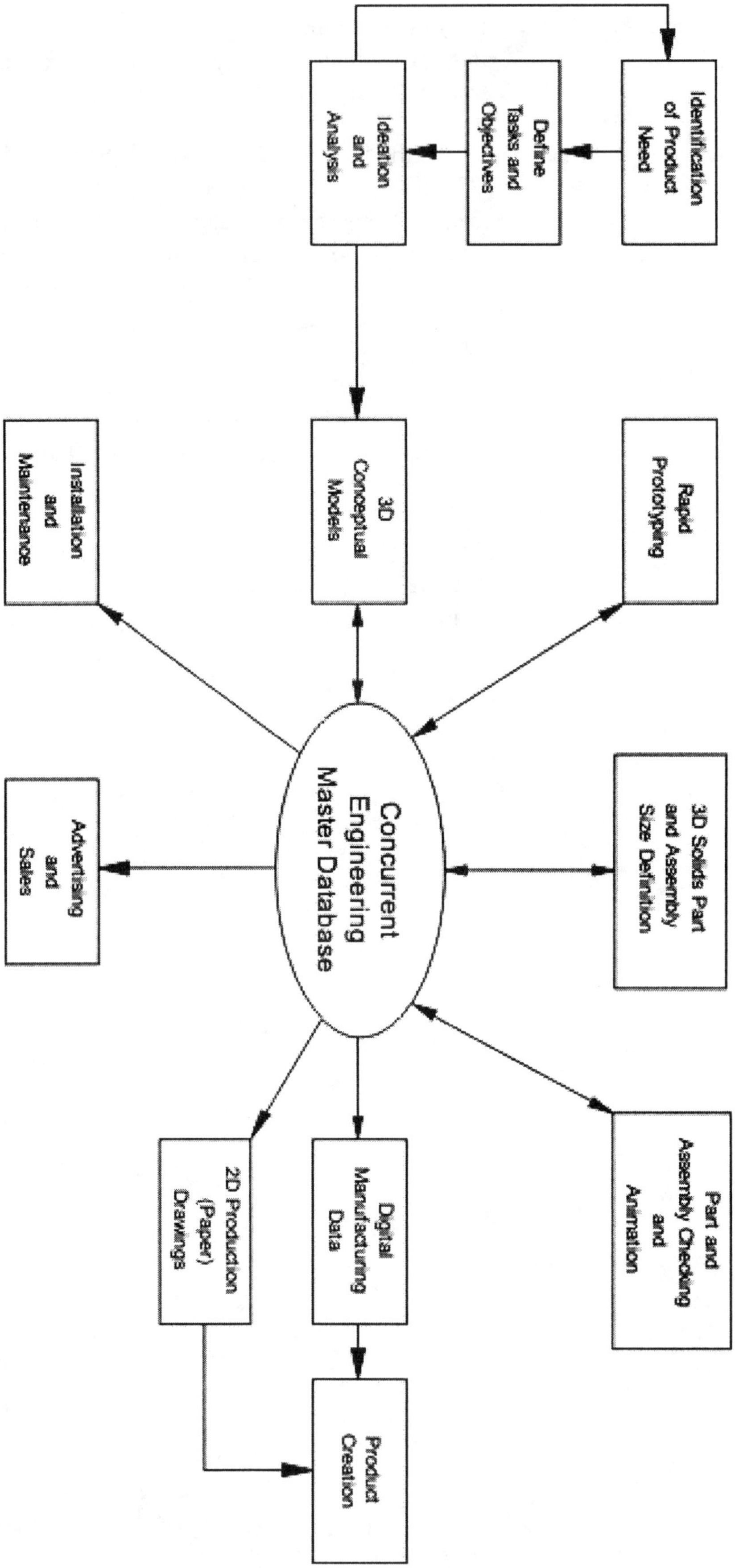
Identification of Product Need
Define Tasks and Objectives
Ideation and Analysis
3D Conceptual Models
Rapid Prototyping
Installation and Maintenance
Concurrent Engineering Master Database
3D Solids Part and Assembly Size Definition
Advertising and Sales
Part and Assembly Checking and Animation
Digital Manufacturing Data
2D Production (Paper) Drawings
Product Creation

Reverse Engineering

Definition by Kathryn A. Ingle in "Reverse Engineering".

"A four stage process in the development of technical data to support the efficient use of capital resources and to increase productivity."

Early use of this design method evolved from the need to service, rebuild or replace factory production machines. These machines were designed and built in starting in the 1880's. Often the machines were hand built by craftsmen or inventors with little documentation. Designs were unique and robust. The machines often operated for many years before significant repairs were needed. By the time repairs were needed, manufacturers, craftsmen and inventors were gone.

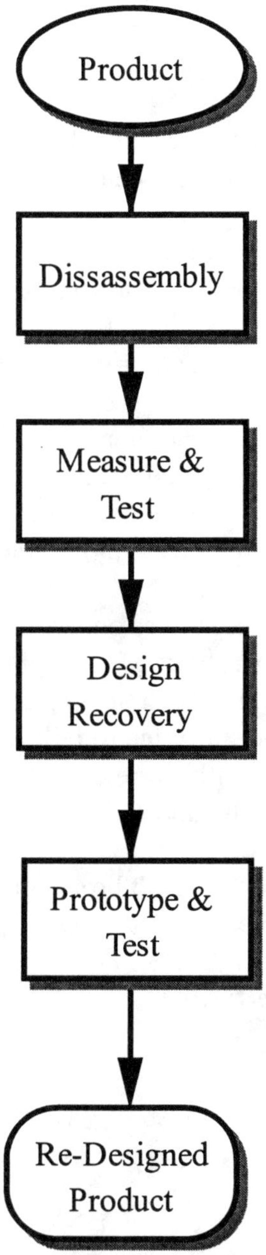

Tearing down the machines, measuring the parts and analyzing the dynamics is a way to recover the original engineering data. There is a strong need to approach the process in a highly organized and professional manner.

Reverse engineering usually focuses on individual components, working from small parts to sub-assemblies and up — working from the inside out. Part operations must be verified. Relations to mating parts are critical , ***recovering the design intention is the objective***. Careful physical size and material measurements are needed. Data is recorded on detail drawings and materials analyses. Once the primary data is complete, strength, heat transfer, dynamic and other data can be implied. Maintenance records, customer complaints, warranty replacements and similar records may supply valuable failure data. Accuracy and completeness of data is extremely important. Any re-design based on incomplete information would defeat the process.

Comparisons between actual performance and needed performance can be conducted to optimize a re-design. Each part must be better suited for its function; more efficient, better quality and lower in cost to justify the process. "The Reverse engineering process strengthens the weak links in any system", states Kathryn Ingle. "New documentation support and improved system maintenance are important by-products."

An interesting case of reverse engineering is the production of computer chips. INTEL corporation developed the 80386 and 80486 chips which drive most of the "IBM type" personal computers. CYRIX and AMD corporations decided to produce competing products. Instruction sets for the INTEL devices were public record. It would be highly illegal to copy the internal computer codes of the INTEL chips - INTEL would sue very quickly. However, by working within very carefully controlled conditions, programmers could construct their own ***original*** versions of the existing functions.

Reverse Engineering is neither patent infringement nor theft.

This example illustrates the fine line between reverse engineering and patent infringement or outright theft. Any patented component of a larger system cannot be a candidate for the process. Engineers may invent another unique component which replaces the patented item but no one may take apart and outright copy the protected item. "Design infringement is not the intent of reverse engineering. Technical design documentation for maintenance and supply support is the desired end."

Companies may conduct reverse engineering studies on their own products. CHEVROLET ran an advertisement which showed a team of workers disassembling a truck selected straight off the production line. The goal of the exercise was to evaluate assembled tolerances, weld reliability, assembly methods, etc. Similar studies are conducted by each large company on competing products. Foreign countries studied American automobiles in the 1960's and often produced superior, more efficient, lower cost, safer and more desirable automobiles in the 70's and 80's. American companies suddenly had a very large gap to overcome and lost much business. Designs were not copied — they were ***improved***.

Aircraft companies may conduct similar studies on airplanes that have flown for many years. Older models are brought back for all kinds of structural, metallurgical, vibration, and flex tests. Ultimately the structures may be subjected to failure loads to find out just how strong they really were. Design practices and strength calculations can be updated.

A popular motto a few years ago was "If it ain't broke, don't fix it." This thinking can lead to obsolescence.
A better statement for the future might be ....
Even if it ain't broke — make it better.

GOOP

GOOP is a registered trademark name. It is a hand soap used by auto mechanics and other people (even engineers) who need to clean grease, paint or similar materials from hands or clothes.

Need defined. Traditional methods of cleaning grease from hands involved sharp abrasives and strong soaps. These materials often left the hands sore and very dry with repeated use. Products included BORAXO and LAVA soap. One company marketed a gel type cleaner that softened grease. The gel-grease mixture had to be wiped clean on absorbent towels. The gel turned very sticky and gummy in water.

Design. An engineer who, as part of a re-design project, often got very greasy became very dis-satisfied with the hand cleaning process. He visualized a hand cleaner that would soften grease, wash off with water and leave hands in better condition.

Refinement. Working part time on the problem, the engineer researched a number of formulas from handbooks and cooked up many trial batches. This process took four years to perfect a workable product. Many formulations did the job but proved too expensive or unstable for mass-market use. Chemical skin irritations and dermatitis had to be minimized. Friends were called upon for tests.

Market research. One great concern was that large soap companies would seize the idea and market similar products. Trips to the library and other research turned up statistics:

- 25% of people get hands dirty daily.
- 33% of people get hands dirty once a year.
- Sales of 5,000,000 cans of hand cleaner per year if only 10% of people would buy a can.
- Large companies would only be interested if the sales would be around $100,000,000 per year.

Decision ..... Safe to proceed. The inventor felt that; (1) there was enough market to proceed, $15,000,000 to $25,000,000 per year was a nice income for a small startup company, and (2) big soap companies would not interfere.

Getting to market.

How do you get a new product to market when you have very little time, money, advertising budget and no consumer recognition? The hard way. You load your old station wagon with cases of product and hit the road. You give away cans to automobile repair shops. You talk auto-supply stores into putting a few cans on the counter. You approach the big super markets and they turn you down - so you talk the small *mom and pop* stores into putting a few cans on the shelf.

Good new products will find the market. It takes exemplary patience, incredible hard work and persistence but, the job can be done.

Patents are not always advisable ...

"I never did patent GOOP", said the engineer/inventor. "I have a number of patents so I know the process and the pitfalls. A good chemist could have read the disclosure and cloned my formula."

"As it is, GOOP is still a market leader, a very capable, safe product which gets the job done as well as anything to come along. It cleans very well — and, it leaves your hands soft."

CONTROLLER FOR A DISABLED PERSON

Disabled people often have great difficulty performing simple tasks. Depending on the disability, turning a light on or off, or tuning or setting the volume on a radio may be impossible. People like to be able to control these things. No wants to have to ask someone else to do a simple job.

Consider Chris - at the age of twelve he was shot in the spine as he and a friend played with a handgun. He is totally paralyzed with no leg or precise hand movement. How can a safe, reliable and inexpensive interface be designed so he can tune a TV or set the volume from bed? How will Chris be able to interface with such a control?

Defining the problem.

1. Control radio, TV, disk player, tape deck, VCR, lights, fan, heater, answer telephone without hands or feet.

2. Control must be safe, simple, inexpensive, reliable, easily repaired.

3. Input cannot be physically activated.

4. The common element in the system is that all the devices are electrical.

Input options

- Voice
- Head movement
- Eye movement
- Breath

Control options.

1. Modify devices for direct wired connection.

2. Use infrared hand-controller type signal.

Safety Considerations.

While direct wire connection to a radio or t.v is possible, it has two drawbacks. First, it could be dangerous due to stray voltages. Some devices use a "floating ground" which can be lethal in the event of an internal malfunction. Second, direct connection involves getting inside the device and cutting wires. This probably voids any warranty and makes replacement of the unit more difficult.

Expanding Infrared Controller Capabilities:

Infrared control of appliances is a very common method of external connection. Radios, TV's, tape decks and VCR's provide this function at little extra cost. How can lights or fans or a telephone be controlled by infrared? Research turned up an interesting system called X-10. This family of devices uses existing house wiring to carry signals to modules which are plugged in to wall sockets. By imposing a high frequency pulse type code on the regular house wires, the control modules can sense whether to turn on or off. By plugging a light, fan, air conditioner, heater or other common appliance into the module, that device can be controlled. Even non-house voltage devices like a telephone or thermostat can be controlled by using an isolation type module. Universal hand type infrared controllers are available which can learn codes from other controllers and thus control a multitude of devices from various manufacturers in one unit.

Interfacing to the person.

Now, there is a common control system on the output side. We still have to interface to the person. This side of the system must be non-restrictive, safe, inexpensive. Head movement interfaces usually require the wearing of some type of head band which can reflect light. This is somewhat unpleasant and confining. Voice-input systems are very expensive, require fast computers, and are not very reliable. Eye movement sensors, like voice systems are very expensive. Breath control can provide positive signals at very low cost.

Chance remark opens a door ....

In discussing the input problem, an engineer mentioned a spec sheet he had received for a low pressure switch. The switch cost was only a few dollars and was designed to sense liquid levels through a low pressure tube. Samples of the switch were tested and it was determined that slight pressure changes activated through a soda-straw would toggle the electrical contacts. Puff and one set of contacts would close — sip and the other set of contacts were activated.

First design.

This unit was designed using relays. Potter and Brumfield Company heard of Chris's plight and donated a number of relays including a big multiple step type. The stepper relay was wired to one side of the breath switch. By puffing repeatedly, Chris could select any of sixteen functions. He would then sip to activate a particular function. The activation relays were hard-wired to a universal infrared controller. This was the only device which had to be modified. A set of lights on the front of the relay box showed which function was currently selected.

Second Design. Solid state circuits (chips) replaced relays. A design similar to the first was soldered together using 74xx series counters, buffers and switches. The parts cost was only a few dollars. This unit was quieter, smaller, very rugged and required much less maintenance. Its demise was at the hands of a well-meaning technician who replaced a power supply but, wired a plug backward.

Computer controller. This unit was developed after a new infrared hand controller came on the market which could be activated from the serial port of a computer. Earlier designs required soldering very fine wires to an existing infrared controller. Much greater flexibility on device control and command sequencing is possible with a computer. Software can take over many of the functions previously set in hardware. This is an excellent retirement job for an older computer. Input is through the game port — the puff and sip switch is wired to the "fire" buttons.

Voice control is probably the next step. Experiments with voice input have not been very satisfactory. The cost of this technology is coming down so, the design effort continues ....

Oh, about Chris. He is now married, a Boy Scout leader, has two college degrees, a singer, teacher and actor. He recently "played a dead guy" in the play Our Town.

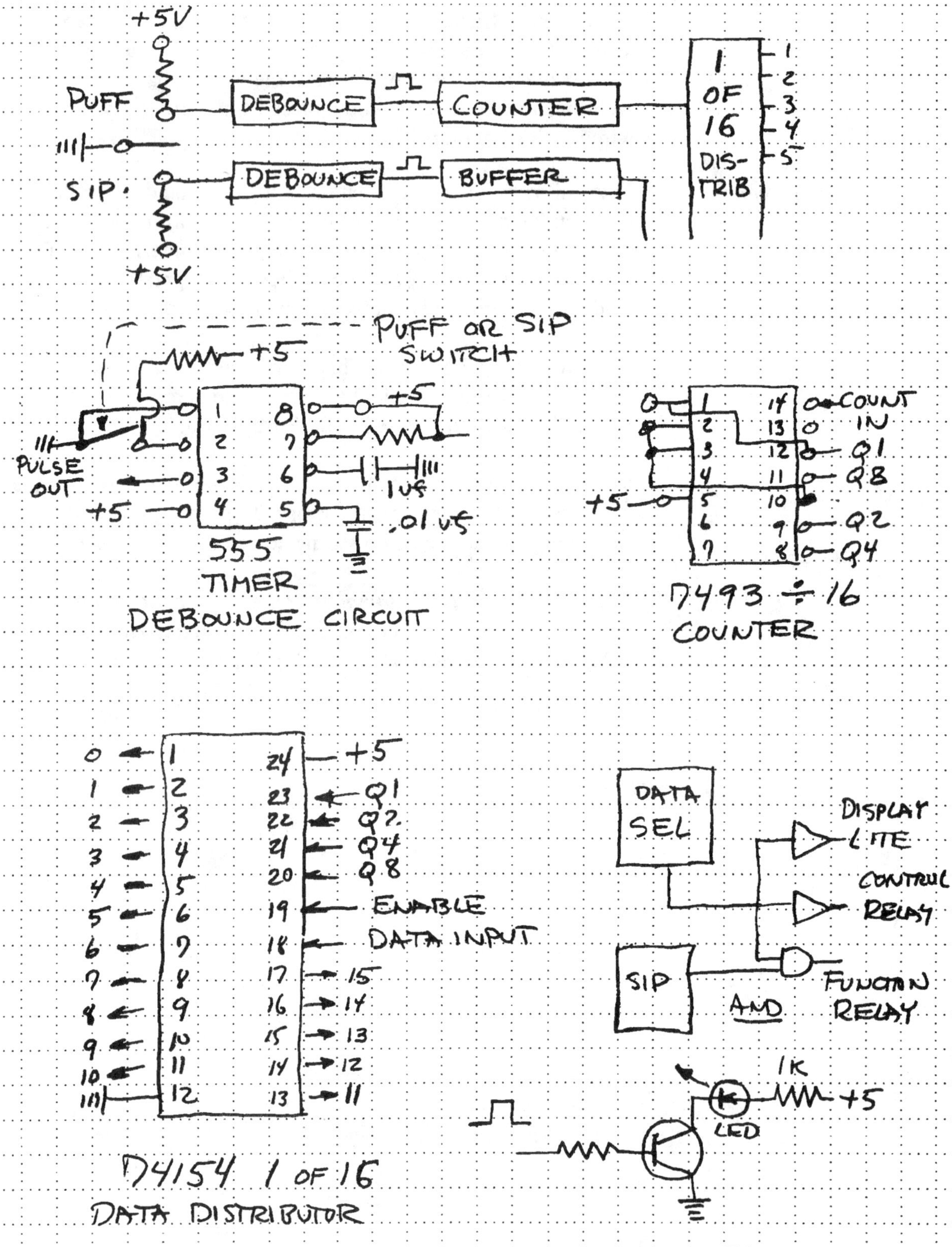

Design Sketches for 2nd Controller

Fast Food - McChicken Sandwich and Chicken McNuggets (tm McDonalds Inc.)

One would not normally think of an engineer being involved in the development of a fast food. The ***need*** in this case arose from an effort to maximize production line efficiency. Imagine a plant which fries *sixteen tons* of chicken per hour, twenty four hours per day. Most of this product is sold as frozen, ready to heat TV dinner meals.

Seasonal demand reduces production efficiency.

The problem with this type of product is that the demand is seasonal, causing more sales at certain times of the year. Maintaining optimum production is essential due to the highly competitive sales market. Processing, frying, packaging with other foods and flash freezing involves close coordination. In fact, the entire operation from egg to chicken to dinner is highly coordinated. Prepared meals cannot be stored frozen for many months because the oils in the food become rancid.

In a quiet moment, the chief engineer for the plant envisioned a "chicken hamburger" type food which would be formed from chopped pieces of meat. Removing the bones would reduce the freezing and storage requirements. The product would be flash frozen in raw form, eliminating the rancid oil problem.

February, 1977

Experimenting with formulations, the best tasting product was composed of dark meat, white meat and skin. This patty was moist and appeared to be just like a boneless white meat fillet. Another problem comes from cooking. Moisture in the food is driven out in the form of steam. This is a total loss and represents a significant cents-per-pound amount. The engineer developed a marinade which would seal the meat to keep the moisture in. The marinade also provided a flavor for the product. A light batter was applied with additional spices.

October 1977. Bingo!

By chance, McDonalds restaurant was looking for a couple of chicken meals to enhance their popular beef menu. Executives from the main office attended a tasting. They were enthusiastic about the product but were not happy that the skin was included in the patty. Further tests showed that the patty without the skin was too dry and lacked the taste of the original. The skin went back in.

McDonalds test marketed in Pennsylvania and Minnesota. Customers were enthusiastic. The new product was an instant hit.

More research brings forth another hit.

Having a family of small children, the engineer was aware of how much kids like to eat with their fingers. Using the same meat mixture the engineer compressed bite- size pieces. *Chicken McNuggets* became a favorite meal for many young children.

Fast freezing of TV dinners was done in long freezers where the product moved slowly through on long conveyor belts. This setup was very inefficient for the chicken patties and nuggets. The engineer had worked with vertical ovens for cooking pastry and investigated the concept for freezing. Vertical freezers were developed where the product starts at the bottom and spirals to the top. This design handles a large volume of product in a very small footprint. Patties and nuggets were flash frozen using liquid CO_2 or nitrogen.

A need to keep a highly integrated process at optimum production led to the development of two very successful products.

Note: The current meat content of the items depicted may have changed. The original contract was fulfilled and the contract for producing the items went to another company.

Software Engineering

Over the past quarter century, the growth of micro electronics and computer systems has been enormous and the impact on the way in which human affairs are conducted has been greatly impacted by this explosion.WID computer systems.

New Technology Spawns New Jobs.

Attendant with the widespread use of these devices has come a variety of occupations little known or completely absent from fields of human endeavor of the past. Some examples include digital design, semiconductor fabrication, and micro electronic engineering. In addition, the foundation of this industry has been comprised largely of the classical job functions associated with the processes of product design and manufacture. Designers, managers and supervisors, documentation developers, marketing specialists, book keepers and auditors, and assembly line workers have all played key roles in the success of the revolution.

Computer Programming follows Engineering Design Process ...

One of the occupations arising from the evolution and growth in computer utilization was the programmer. His products, the computer program, share many of the characteristics associated with the processes for design and manufacture of a mechanical component such as the fuel injection system of an automobile or an engine of an aircraft. The programs may be very small in size containing code to provide a very narrow range of function or they may be quite complex having many components providing a broad spectrum of functionality. They may regulate the operation of mechanical systems or serve the commercial world in accounting and inventory control. In general, they provide comprehensive symbol manipulating capabilities that are key to activities as diverse as mathematics, composition, graphics, music and many other forms of communication each having its own unique set of symbols and rules for their combination.

Problem Background and Definition.

With this background, the study presented here will focus on a large software program that has been designed to run on a mainframe computer located at a data processing facility in the United States and which serves a number of user facilities throughout the world. The program provides a variety of finance and accounting applications for its terminal users that include payroll entry, travel expenditures, training schedules, wages, taxes, and several other basic components that form the suite of a personnel payroll accounting system for a large organization.

This application was developed and fielded in the late 1980's and from the user point of view, performed satisfactorily according to the specifications developed for the program design. Input to these specifications came from users, managers, and accountants working in coordination with programmers, training and documentation staff, computer operators, and because of the wide geographic distribution of users, communication experts as well. The development of this software package followed standard practices generally applied to the creation of substantial sized applications for large mainframe computers and the processes employed could equally well describe any of a great number of development projects of similar type and scale.

Since industry has been in the software business for a considerable period of time, these standard practices have been refined to the degree that the activity is commonly referred to as "software engineering" thereby placing it in a highly structured and rigorous context and suggesting the similarity that certain aspects of the programming discipline share with engineering methodologies.

Following the usual period of startup problems and resolution that are to be expected from any large programming effort, the application was accepted by the customer, management, and contracting officials and was put into full scale production use. After some operational experience had been gained, it was evident to performance analysts, those people responsible to keep mainframe computers working at peak efficiency, that some of the routines were extremely resource intensive and required a great deal of time to execute a specific set of transaction classes.

Need Defined. This was an intolerable situation from the analyst's viewpoint. Since the programs consumed such large amounts of processor time and required millions of operations on disk drives the application limited the number of users that could be accommodated by the computer system in a responsive manner. To those using the system, waits were experienced for transaction completions which they came to view as normal and, therefore, did not bring to the attention of persons who might have addressed the issue.

Multi-user operation. A brief digression should be made at this point to clarify aspects of this process that may not be evident to the reader unfamiliar with the internal workings of computer systems and the programmer's role. First, any computer system is a finite resource containing a central processor, memory, and storage devices such as disk and tape drives. Each user tapping into a multitasking system consumes some part of these resources and the amount required for each person depends, to in large measure, on the programs being used. If the programs are inefficient, they will require more resource than those which produce the same amount of work with smaller demand. Therefore, more efficient programs allow more users on the system at one time than those with larger requirements and provide greater responsiveness to all users.

One might think of a computer system like a telephone booth. The booth can accommodate one person at a time giving the user the full resource just like the personal computer provides its user the full scope of its resources to run programs. If two or more persons wish to use the phone at the same time, a queue must form. In just this way, if more persons try to use a computer system than it can handle, queues are formed internally and waits for the resources are experienced at the terminal.

"Over-design" occurs ...

In the process of developing programs, a general rule of thumb that many software developers follow is "make it right, then make it fast." In other words, be conservative and be sure that the programs will function properly when new code is written. Then, after verifying that everything is satisfactory, one can apply some cleverness to the way in which the program is constructed and improve the overall speed of execution. This can include the structure in which data is stored and retrieved from disk systems and it is there that big gains are made in the performance of computer programs because the disk systems, being mechanical in operation, will generally require magnitudes more time to carry out a basic operation compared to the semiconductor speeds of the memory and central processor.

The degree to which efficiencies are implemented will depend upon a number of factors including the time allowed the programmer to complete the task, the number of opportunities for speed improvement, and the money left in the budget.

Need becomes apparent.

The situation might have continued status quo but for the persistent work of the performance analyst who brought the problem to the attention of developers and managers on several occasions. In each case, the problems were presented in terms of the input and output requests to disk systems and the processor time used by an average transaction. In general, the numbers were quite large and had no impact upon the listeners who had little appreciation for the arcane quantities involved. Consequently, no action was taken to correct the situation.

Dollars and Cents approach gets attention.

The analyst finally got the attention of his audience by casting the resource utilization for a typical transaction in terms of actual cost to the customer and comparing this cost to that of similar transactions processed by similar programs. With these numbers, it became abundantly clear that some of the programs were extremely inefficient in the way in which they utilized computer hardware but more importantly, it was costing a lot of money for this extravagant demand.

Setting Objectives ...

Sanctioned by management, the developer was contacted and the problem outlined with clear documentation identifying the parts of the program that required improvement. This stage contrasts with that of the initial design in that the user is generally not part of this process unlike the key role the user plays initially. An agreement was negotiated for a redesign of the programs and a time table and cost estimates developed. This part of the process would involve managers and financial officers with higher levels of management making the decision that is generally supported by a "business case" outlining the benefits of the expenditures for the work to be done and the consequences should it be denied.

Setting Budgets...

Statements of work and contractual arrangements are also developed at this time and milestones established for the work. Where necessary, changes to the documentation were made and a training budget and instructional programs created to provide any additional training for the users of the program in the areas affected.

This was not necessary in the case considered since the changes to the application would be completely transparent to the user who would only feel a difference in the response time at the terminal.

Testing Phase. Contracts specify performance.

A test program was developed that would first exercise the new software in an isolated non production environment, and when functioning properly, the new software would be migrated to a live production system where further testing would be conducted by a small number of users that would limit the scope of impact caused by any bugs found in the modules. Problems experienced in either of these tests would be cause for the developer to correct those problems and a new cycle of testing would be initiated. Unsatisfactory behavior could arise either because the programs failed the basic data manipulation requirements or because they could not provide the response time specified in the contract agreement. These details must be part of the acceptance criteria agreed upon by developer and customer alike.

One of the most critical aspects of software design is the testing phase where the programs must be given a thorough and realistic examination. Two factors conspire to make this an extremely difficult task. First, the application could be a very large and complex collection of programs having a wide range of dependencies on one another that would make the testing of all combinations of those relationships impossibly large to simulate in a reasonable time. Second, testing will generally be done with a limited number of live users for the reasons outlined above.

Extremely high reliability is required.

Experience has demonstrated that when full scale operation of the software is underway, the likelihood of aggravating catastrophic failures because of the dependencies and stress on resources is magnified. There is no substitute for a thorough testing program that is capable of simulating the real life environment that will be encountered in the production version of the product both in terms of interrelationships and customer populations.

Several iterations of the program development and test phase were implemented and the final version of the software was put into full production service. Still, some small problems have been reported and resolved. The package now has been in service for several months and has remained very stable and robust having survived unexpected situations that have arisen in the environment that it shares with many other applications.

Of course, dealing with the unexpected is quite difficult to accomplish when creating any new system whether hardware or software. However, the designer should always keep in mind the question of "what if" and try to protect the final design from those possibilities. The extent to which protection for hardware and software systems is provided must, of course, be balanced with the cost and time required to implement those features.

For critical applications such as financial and mechanical systems, the extra consideration is well worth the effort and in many cases where bodily harm could result from design failure, essential.

Communication between members of the design group is critical.

Finally, with the wisdom of much experience, one piece of advice is offered. Never, never make any assumptions about the expectations, operation, or relationships to other systems that the design must satisfy. ***Frequently, problems have been created for designers who have made some assumptions about the design and neglected to communicate with other members of the project.*** It is very easy to let intuition take over and lead the process down the wrong path or leave out some important detail that results in costly complications that could be easily avoided. Constant review of the project with all parties involved in the process will short circuit many of the problems that assumptions will create and is strongly encouraged.

These comments were provided by Dr. Robert Arnzen. Bob is a graduate Mechanical Engineer in charge of a world-wide computer network. Bob is blind and has written many complex computer programs.

Centrifugal Casting

Die cast aluminum parts are common. Many times the outer surface of the part is smooth and free of voids. However, the internal metal structure is often not too good with gas pockets, poor crystal structure and poor strength. Clever design of molds and pressure casting may reduce some of the problems.

The problem.

A company, which manufacturers air conditioning compressors, was having problems with aluminum connecting rods breaking. This was causing a number of warranty problems. Customers, installers and sales outlets were unhappy. Manufacturing techniques involved careful inspection of rough castings before expensive machining. Xray, XYGLO and other inspections rejected a number of castings, but some rods got machined and were still defective. The manufacturer issued a bid request and spec sheet for more reliable connecting rods.

Innovative approach to a solution.

Charlie Muench was chief engineer for small, very dynamic company which specialized in high quality metal castings. Mr. Muench realized that the connecting rods would be much stronger if the internal metallic structure were better and if there were very few gas bubbles or other inclusions. He looked for a process that would allow the cast parts to attain higher density with fine internal structure.

As with any design problem, a number of solutions were researched, considered, and discarded. High pressure casting for example, would produce more dense metallic structure but, did not allow gas bubbles to escape. Looking at a wide spectrum of manufacturing process, Charlie took note of centrifugal casting. Up until that time, centrifugal casting had been used to form bronze bearings for machines. The spinning process was ideal for forming the hollow cylindrical shape. No other commercial applications existed.

Why not build a multi-part permanent mold, pour in hot metal and spin the mold until the metal solidified? Would this allow the gas bubbles to "float to the top"? Surely the resulting part would have a higher density and better metallic structure.

An old washing machine facilitates a breakthrough.

Testing took many months. Charlie adapted an old washing machine with a spin cycle to the testing process. "That old machine looked pretty sad. No case. Just the innards sitting there". Small parts were cast. Starting with simple shapes then working up to more complex parts. Careful design and machining of the dies took time. Casting cycles were tested as timing was critical. If complex shapes were not ejected promptly, they might freeze into the dies. Metallurgical studies verified that the internal structure was greatly improved. Gas pockets were virtually eliminated.

"The old washing machine turned too fast. We had to slow it down. "

The real machine takes form.

"We built a full size centrifugal casting machine. I designed the power system using hydraulic motors so we could control the speed easily. I designed two control modes - a manual mode which was activated by a series of push buttons and a automatic mode utilizing Eagle timers. The manual mode allowed us to fine tune each production cycle".

"I overlooked one item on the first design. I assumed that the friction in the drive and bearings would slow the machine. Not so. When you get a ton of weight rotating that fast, it takes a while to come to rest again. I went down to Yates Ford and bought a Thunderbird front disk brake and caliper and installed it on the bearing shaft. We often ran a whole year on one set of brake pads."

Casting Connecting rods and an unexpected setback.

"We machined a die which would cast a number of connecting rods in one cycle. Output was very good - we had very few rejects and the final machining turned out a beautiful product. Our rods were considerably stronger than the current ones. We submitted samples and quoted a price well below the current price. After all, we did not have to inspect each rod nearly as much, we had far fewer rejects."

"**Our bid was rejected — I couldn't believe it**. The purchaser said our bid was too low and we would go out of business. They did not want to risk losing a supplier. I knew we were in good shape. We stood to make a nice profit. They simply did not understand that we had a far superior process."

All's well that ....

"Another contract came along. This was for in-flight fueling nozzles for military aircraft. Our previous experience allowed us to be the only company that could produce an acceptable product. We supplied a lot of those. We also built some outstanding cast aluminum wheels for Mustangs and other cars. Those wheels were incredibly strong."

Many thanks to Charlie Muench for this and other stories related over many miles on the road. Charlie and Mac Yates raced COBRA #7 for many years throughout the mid-west in S.C.C.A sports car competition.

Design Problems

Examples of design projects and sample drawings are shown in the previous pages. Design projects for an introductory graphics course must be limited in scope due to the time available. However, the experience gained in creating sketches, formulating sizes, fits, choosing standard parts and creating engineering drawings is very valuable for future work.

Graphics course time is limited so projects must be reasonably small.

While the future educational goals of many students may be very diverse, few if any courses in the chosen curriculum may be taught in the freshman year. Many students do not choose a specialty until later. This limits the range of possible projects. Curriculum specific projects will be experienced in future courses.

Mechanical type projects incorporate many features which must be considered for any type of design. These include form, function, power, critical fits and clearances, strength, manufactureability, maintainability, cost, etc. While the projects are somewhat small, students will find that there are many decisions to be made. At first it is difficult to focus on exactly what the critical areas of the design might be. Students tend to spend too much time on unimportant details - this is part of the learning cycle.

Start with sketches.

Much of the early design cycle involves sketches. These sketches are quick to create and provide a very clear method of communication. Formal drawings (CAD) will be needed in the refinement, decision and final design phases. Freehand sketches must be very neat, proportional and clear. Designers start by imagining solutions. These mental pictures are captured in the form of sketches which are refined many times before the final solution is chosen.

Use engineering grid. Sketch to scale.

Overall sizes or product limitations are usually known. Set up some type of scale based on the paper size and the product size. For example use 1 grid = 1/4 inch or 1 grid = 10mm for a small part. Use a scale which is easy to convert to real size just by counting grids. Perform the work with the attitude that you will be presenting the sketches to your boss for approval.

Neat sketches convey confidence, knowledge, an orderly disciplined mind, and integrity.

Use detail, sectional, pictorial sketches to communicate clearly.

Where mating parts must exist in the same space, use colored lines to differentiate items. Create pictorial and orthographic sketches, detail and sectional views as needed. All of these drawing techniques exist in order to provide the clearest possible communication. Pictorial views are essential if the person who must approve your design does not read orthographic drawings.

Well thought out sketches are essential for CAD drawing.

Sketches provide essential information when the formal drawings must be made. By working out details like diameters, material thicknesses, major features, assembly methods, etc., computer drawing progresses must faster.

Choose and place dimensions on the sketches first. This will help to locate missing dimensions and will help greatly in the spacing and placement of views on the CAD system.

Make notes on the sketches. Work out limits and fits for tolerance dimensions. Record information about purchased items.

Final Drawing Setup — CAD

Once the preliminary design sketches are approved, the design must be created to exact size on a CAD system.

- Create a standard title block. The size of the title block depends on the plotter and paper size available. Draw the title block full size so it may be plotted 1=1. (Most CAD software allows windows or view ports to be placed in a standard format page so the actual design can be plotted at a reduced scale if needed.)
- Make a list of the organization of the product drawing. CAD software allows the use of blocks, objects or layers in order to control the items in the drawing. Each part should have its own block or layer, and color. Segregating major parts of the drawing allows better control for dimensioning and presentation drawings. Set up this structure on paper.
- Create a master drawing. Draw each part exactly where it will be in the final design. Again, use individual blocks and colors for each part. ***The master drawing will have all the parts in one drawing.***
- Detail part drawings may be created by loading each object from the master drawing into window of a separate title block. Dimensions will be shown in the detail drawing.
- Assembly drawing. Several types of assembly drawings may be created. Pictorial, exploded pictorial, sectional, etc. are relatively easy to format from the master CAD drawing. Exploded drawings are made by creating overlapping windows, loading the master drawing into each window then

turning off all but the item to be seen. Some CAD software allows individual parts to be loaded into each window, thus simplifying the process.

Other interesting drawings (time permitting) include:
- Shaded or rendered pictorial drawings for advertising.
- Stereo pictorial drawings.
- Animation sequences.
- CAD/CAM drawings.
- Automated parts lists.

Projects

These suggested projects are at the "Preliminary - Tentative Design Phase". That is, the need has been verified and some preliminary design work has been done. Critical dimensions have been established.

Create sketches and final drawings.

Project 1. Caster Wheel Assembly

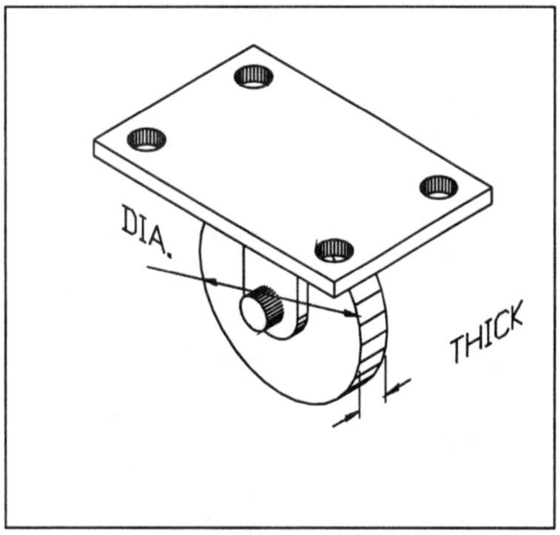

CASTER WHEEL ASSEMBLY
Each student or group will be assigned a different value for diameter and thickness.

___Design a caster assembly.
___Minimize the overall height.
___Improve the appearance.
___Improve the strength.
___Design a similar item with a steerable wheel.
___Design a brake for the wheel.

Check items assigned.

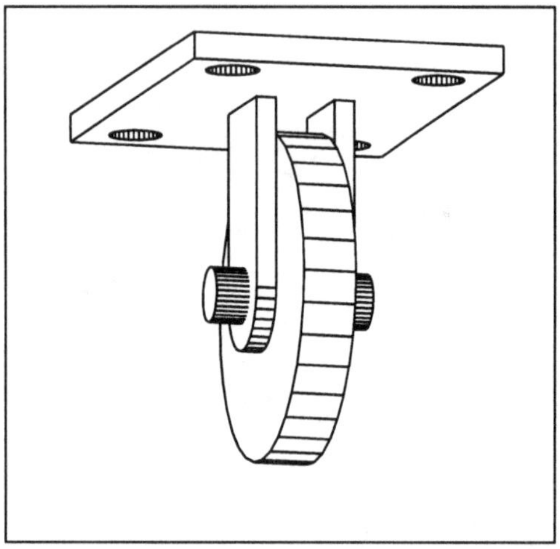

Concept drawings are shown.

Project #2 Peristalic Pump

Each student or group should be assigned a different diameter for the center line of the tubing and a different diameter for the tubing.

This type of pump is used in hospitals or food processing to avoid contamination of the fluid in the tube.

- Improve the appearance
- Prevent the tubing from slipping.
- Prevent the tubing from falling out.
- Calculate (rough estimate) the volume of fluid pumped at 200 RPM.
- Option - Vary the pressure on the tubing by the pinch rollers.

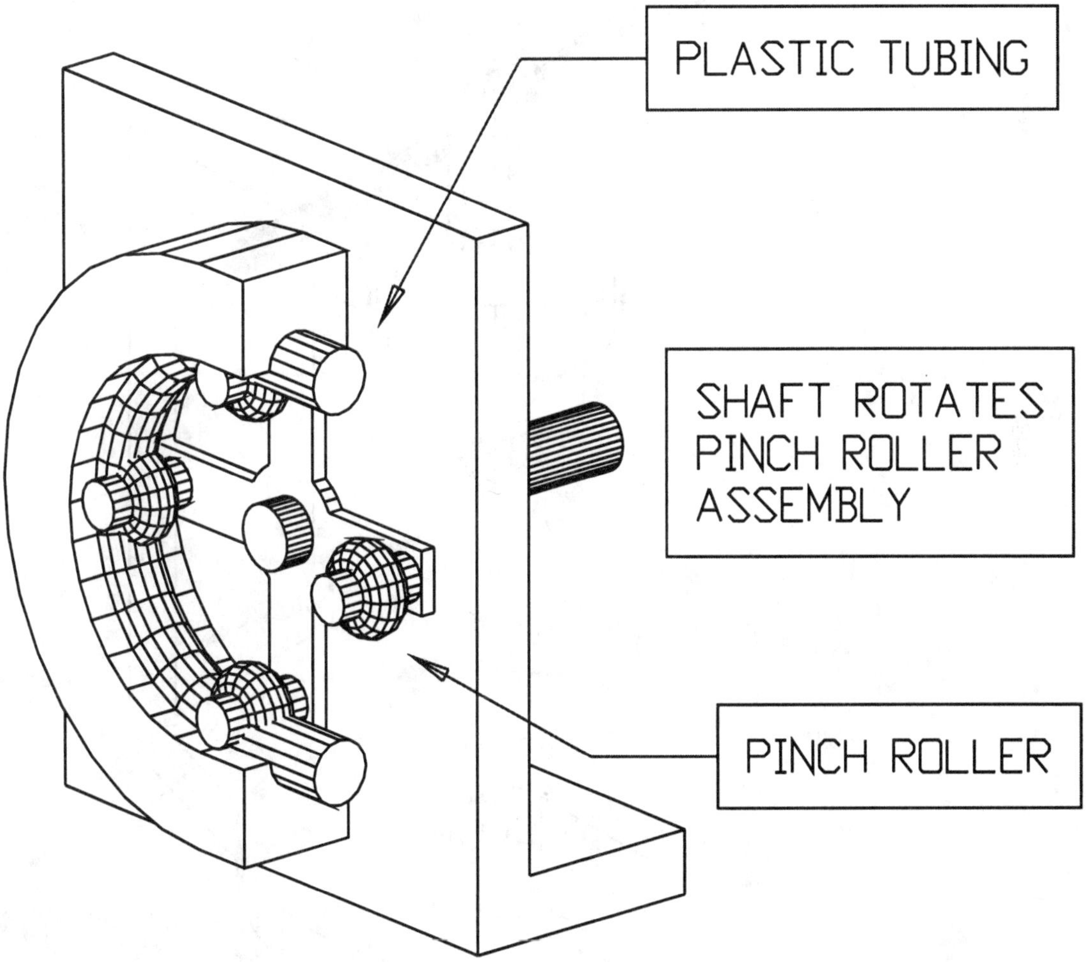

Peristalic Pump Concept Sketch

Project #3 Screw Press

- Design a press to force the bearing into the wheel as shown on detail drawing page F-15 of this booklet.
- Design the tools to hold the wheel and the bronze bearing during the assembly operation.
- What other power systems could be used?

A concept sketch is shown. The frame should be stronger. The frame should be held down to a workbench due to the turning force required.

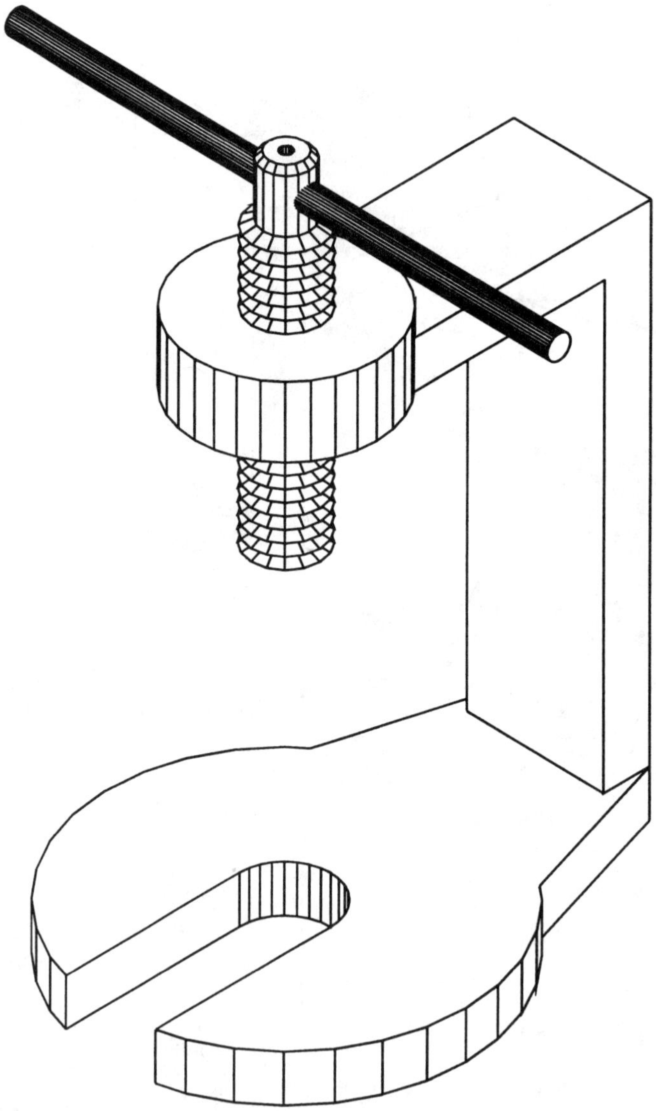

Other Short Projects:

Obtain samples or examine these common items. Other Short Projects

Obtain samples or examine these common items.

- Centrifugal pump {water pump on car}
- Stepper motor {used on disk drive in computer}
  Be very careful of samples of these motors as they have extremely powerful magnets which can damage floppy disks if brought too close.
  Study the construction of these motors. Re-design for 60, 100 or 360 steps per revolution.
- Design a brake for a grocery shopping cart. {Rolling carts do a lot of minor damage to cars in a parking lot}.
- Design a heavy duty cart for hauling heavy loads around a garden shop or builder's supply. Cart must have positive brakes which are always applied except when movement is needed.
- Design an educational "toy" which would demonstrate the mechanical advantage of various pulley systems.
- Design an aluminum can crusher for use in the kitchen.
- Design a nut-cracker using the principle of the toggle mechanism. (Examine a pair of VISE GRIP (tm) pliars. Search for Toggle Mechanism on the internet).
- Design a clamp that can be activated by one hand.
  (Hold pieces in one hand and apply clamp with the other).
- Design a tool to thread a needle for those who cannot see small detail.

Longer Projects

Light pens were once used for input on computers. The photo cell inside the pen synchronized with the pulse which traces the electron beam across the display tube. This allowed the pen to "pick" any place on the tube. Just point at a place on the tube and click the pen switch. Almost all video controller boards still support a **pen** input command.

Why are pens not used now? (The answer lies in the original computer configuration). They would seem to emulate a pencil and make drawing easier. They would be more positive than a mouse and the pen would not slip.

- Design computer station furniture for comfortable light pen use.

- Design a self-contained electric lawn mower. Pollution laws in California now prohibit gasoline powered mowers and more locations are subject to similar federal rules.

- The Gas Company in St. Louis County Missouri asked us to design a gas meter which could be read by plugging a device to a connector box at each house. The read-device would power the gas meter momentarily and digitally record the volume used. No power is available except at read time. The gas meter would have to store the volume used mechanically.

- Carry the gas meter problem forward one step. How could the Gas Company read the meter over the telephone each month?

- Design a mechanical dynomometer (pump, fan, prony brake, etc.) for bench testing lawn mower engines. Horizontal and/or vertical shaft.

- Carry the dynomometer design one step further and analyze the exhaust gas emissions caused by 2 cycle and 4 cycle lawn mover engines. Could you use existing automobile computers and emission sensors to reduce cost?

- Design a shelter system which could be used to house several people in event of a flood, earthquake or other emergency. The system should be portable, store in a small space, be re-usable, sanitary (washable), provide light, heat and ventillation.